RAND NATIONAL DEFENSE RESEARCH INSTITUTE

Military and Civilian Pay Levels, Trends, and Recruit Quality

James Hosek, Beth J. Asch, Michael G. Mattock, Troy D. Smith

Prepared for the Office of the Secretary of Defense

For more information on this publication, visit www.rand.org/t/RR2396

Library of Congress Cataloging-in-Publication Data is available for this publication.
ISBN: 978-1-9774-0166-3

Published by the RAND Corporation, Santa Monica, Calif.

Cover images by Yozayo/iStock/Getty Images Plus and Bamlou/DigitalVision Vectors.

Preface

Force capability depends on military compensation being sufficient to attract and retain the number and quality of personnel that the services need. If pay is inadequate, personnel shortages can result and hurt readiness. The U.S. Department of Defense undertakes periodic reviews of military compensation. Quadrennial reviews of military compensation, in particular, address many facets of compensation related to basic pay, allowances, special and incentive pays, retirement pay, health benefits, and more. A key topic in every quadrennial review is where current compensation stands relative to civilian pay for workers with comparable ages, education levels, and labor-force participation. Although it is not part of a quadrennial review, the present report focuses on military pay for active-component personnel relative to civilian pay. It was motivated by a recommendation of the Ninth Quadrennial Review of Military Compensation—namely, that military pay for active-component enlisted personnel be at about the 70th percentile of civilian pay for full-time workers with some college and that military pay for active-component officers be at about the 70th percentile of civilian pay for full-time workers with four or more years of college. The research reported here asked where active-component pay stands relative to civilian pay in 2016; whether this differs from where it stood in 2009, when the 11th review was conducted; and whether changes in entry-level military pay have been associated with changes in the quality of recruits into the active components.

This document should interest the defense manpower policy community and officials with responsibility for military-pay policy. This research was sponsored by the Office of the Deputy Assistant Secretary of Defense for Military Personnel Policy and conducted within the Forces and Resources Policy Center of the RAND National Defense Research Institute, a federally funded research and development center sponsored by the Office of the Secretary of Defense, the Joint Staff, the Unified Combatant Commands, the Navy, the Marine Corps, the defense agencies, and the defense Intelligence Community.

For more information on the RAND Forces and Resources Policy Center, see www.rand.org/nsrd/ndri/centers/frp or contact the director (contact information is provided on the webpage).

Contents

Figures

Tables

Summary

In the all-volunteer military, pay is one of the most important policy tools for recruiting and retaining personnel. Military pay must be high enough to attract and retain the personnel needed to meet requirements, and one measure of pay adequacy is how it compares to the pay of civilians with similar characteristics.

The Ninth Quadrennial Review of Military Compensation (QRMC) concluded, "Pay at around the 70th percentile of comparably educated civilians has been necessary to enable the military to recruit and retain the quantity and quality of personnel it requires" (Office of the Under Secretary of Defense for Personnel and Readiness, 2002, p. xxiii). The 9th QRMC focused on active-component personnel, and it measured military pay by regular military compensation (RMC), a general-purpose measure of cash compensation. RMC is defined as the sum of basic pay, basic allowance for housing, basic allowance for subsistence, and the federal tax advantage resulting from these allowances not being taxed. The wording "comparably educated civilians" was critically important. The 9th QRMC reported that the RMC profile for enlisted personnel was at the 70th percentile of annual earnings of full-year male workers with high school degrees but at only around the 50th percentile for those with some college education. Because Status of Forces Survey data indicated that, after several years of service, most enlisted personnel had some college and many had completed associate's and even baccalaureate degrees, the QRMC believed that the appropriate comparison was with those with at least some college. It therefore recommended pay increases that would bring the enlisted

RMC profile up to the 70th percentile of wages of civilian men with some college education. Also, the officer profile was above the 70th percentile for male college graduates but below that percentile for men with advanced degrees. Again, because most officers had more than a college degree, the 9th QRMC argued for policies that would increase the officer RMC profile over time to the 70th percentile of the appropriate comparison group.

The 11th QRMC looked at military pay in 2009, ten years after the 9th QRMC made its comparisons (Office of the Under Secretary of Defense for Personnel and Readiness, 2012a). Averaging enlisted and officer education levels and including both men and women in the pay comparisons, the 11th QRMC found that RMC for active-component personnel was at about the 90th percentile for enlisted members and 83rd percentile for officers. The Office of the Secretary of Defense requested that RAND National Defense Research Institute analysts conduct similar comparisons for 2016.

The research summarized in this report addresses three questions:

- How does military pay for active-component personnel in 2016 compare to civilian pay, and, in particular, is military pay above the 70th percentile of civilian pay, the benchmark recommended by the 9th QRMC?
- How do the results for 2016 compare to those for 2009, the year studied by the 11th QRMC?
- Given that military pay relative to civilian pay has increased since 1999 when the 9th QRMC benchmarked military pay, has there also been an increase in recruit quality?[1]

In addressing the first two questions, we used data from the U.S. Department of Defense's (DoD's) *Selected Military Compensation Tables* (Directorate of Compensation, 2017), known as the Greenbook, and from Active Duty Pay Files provided by the Defense Manpower

---

[1] When we talk about military pay increasing relative to civilian pay, we mean that military pay reaches a higher percentile of civilian pay. If RMC is at the 80th percentile of civilian pay, for instance, 80 percent of civilian-wage earners have a lower wage than RMC and 20 percent have higher.

Data Center, together with data from the March supplements to the Current Population Survey, to compare military pay to the pay of civilians with similar characteristics. We used data from the center's August 2009 and September 2016 Status of Forces Surveys on the education distribution of enlisted personnel and officers, and we used data from *2015 Demographics: Profile of the Military Community* (Office of the Deputy Assistant Secretary of Defense for Military Community and Family Policy, 2016) on the gender mix in the military. We weighted civilian workers by the military gender mix then computed a civilian-wage distribution for each level of education. Treating RMC as though it were a wage, we found its placement in the distribution (i.e., we determined its percentile). We computed RMC percentiles for officers and enlisted by year of service, as well as an overall RMC percentile, for 2016 and 2009. Because the military education distribution has increased toward higher education levels over time, we also performed counterfactual calculations (e.g., estimating the 2009 RMC percentile subject to the 2016 military education distribution). In addition, we did computations to examine how RMC percentile has changed over time for specific age and education groups.

For the third question, we estimated regression models to determine the relationship between recruiting outcomes and the ratio of RMC to the median civilian wage of high school completers ages 18 to 22, controlling for other variables. We estimated separate models by branch of service and used two types of recruiting outcomes. Both of the outcomes are for non–prior service accessions. The outcomes are the recruiting rate and the share of accessions who are not high school diploma graduates (HSDGs). We calculated the outcomes for Armed Forces Qualification Test score categories I, II, IIIA, and IIIB. For instance, the category II recruiting rate in a year is the ratio of HSDG accessions in category II to the population of high school completers in that category net of those going on to complete four or more years of college. The share of non-HSDG accessions in category II is the ratio of non-HSDG accessions in category II to the total number of accessions in that category (HSDG and non-HSDG).

The Regular Military Compensation Percentile for 2016 Was Above the 70th Percentile

We found that, in 2016, the overall RMC percentile—taking a weighted average across education levels based on the military education distribution—was at the 84th percentile for enlisted personnel and the 77th percentile for officers. We also computed RMC percentile by level of education, and, in particular, we found that RMC for enlisted members was at the 87th percentile on average for those with some college and the 85th percentile for those with associate's degrees. These findings address the first question, and we can conclude that RMC was above the 70th percentile recommended in the 9th QRMC. In addition, at the beginning of an enlisted or officer career, the overall RMC percentile was often between the 85th and 90th percentiles of civilian pay but was lower at high numbers of years of service, reflecting higher levels of education and consequently higher civilian wages. For officers, RMC was at the 86th percentile for those with bachelor's degrees and the 70th percentile for those with master's degrees or higher.

Tables of the military education distribution show that personnel in higher grades have higher educational attainment, more so in 2009 than in 1999, and more so in 2016 than in 2009. The opportunity to gain further education while in service adds to the value of a military career. The increase in education is consistent with the military's emphasis on professional military education, education as a factor in promotion, and the provision of access to higher education through service-related educational institutions, such as the National Defense University and Air Force Institute of Technology.

The Regular Military Compensation Percentile Was About the Same in 2016 as in 2009

We found that the overall RMC percentiles for 2016 for enlisted personnel and officers were virtually the same as for 2009. This finding uses five levels of education for enlisted (high school, some college, associate's degree, bachelor's degree, and master's degree or higher), and it uses two levels for officers (bachelor's degree and master's degree or higher). When we limited the education levels to the levels used for the 11th QRMC, which were high school, some college, and associate's degree, RMC for enlisted is at the 88th percentile in 2016, versus the 86th percentile in 2009. This is somewhat less than the 11th QRMC's 90th percentile, and the differences come from differences in methodology. The 11th QRMC's estimate assumes that the enlisted education distribution is the same as the civilian education distribution, while we used data on the enlisted education distribution, and there are differences in the way the number of years of labor-force experience is imputed that would lead the 11th QRMC's RMC percentile estimate to be higher than our estimate.

Our RMC percentile for officers—the 77th percentile for both 2016 and 2009—is below the 11th QRMC's estimate for 2009, which was the 83rd percentile. We used the same education categories as the 11th QRMC, although, again, we used data on the military education distribution, while the 11th QRMC used the civilian education distribution, and there are differences in imputing years of labor-force experience.

Our finding that the RMC percentile is the same for 2009 and 2016 contrasts with estimates of the "pay gap" that compare changes in basic pay and the Employment Cost Index (ECI). The changes in the ECI and basic pay since 2010 indicate a 5-percentage-point pay gap by 2016. In Appendix C, we present estimates of the pay gap using basic pay and the ECI and critique that method.

We also compared RMC to civilian wages from 2000 to 2016 for selected age and education groups. These comparisons show a steady increase in RMC relative to civilian pay from 2000 to 2010 and a leveling off afterward. Civilian wages adjusted for inflation trend down from 2000 to 2013, although they have tended to increase since 2013.

Recruit Quality Rose in Three Services as Military Pay Increased Relative to Civilian Pay

Regression estimates indicated a positive association between recruit quality and the ratio of RMC to the civilian wage for the Navy, Marine Corps, and Air Force but not for the Army. The Marine Corps and Air Force increased quality by increasing the recruiting rate for category I and II, while the Navy sharply decreased the recruiting rate for the lower quality category, IIIB. The Army decreased the recruiting rate for category IIIB, as well as for II and IIIA.

Further, the Army had a positive association between the share of accessions that were non-HSDGs and the ratio of RMC to the civilian wage in every category. This was also true for the Marine Corps. These services took more non-HSDGs as military pay rose, other things equal. The Navy increased the share of non-HSDGs in categories I and II but not IIIA or IIIB. The Air Force decreased the share of non-HSDGs in categories I and II and increased the share in categories IIIA and IIIB.

The reason for the Army's different result is an open question. Possibly, Army recruiting became more difficult during the 2000s because of extensive deployments in support of operations in Iraq and Afghanistan, and the Army did not program enough recruiting resources to match the increased difficulty. Possibly, the Army set its recruiting quality goals to hold recruit quality constant as RMC increased. If so, this might reflect an implicit calculation that the marginal cost of high-quality recruits rose relative to that of non–high-quality recruits, and a decision to hold quality near the DoD quality benchmarks of at least 90 percent tier 1 recruits and at least 60 percent from categories I through IIIA rather than allocate more resources to recruiting.[2] The majority of tier 1 recruits are HSDGs (Office of the Under Secretary of Defense for Personnel and Readiness, 2011). We explore these issues in a theoretical model giving conditions for the optimal allocation of recruiting resources subject to two goals: one for the number

---

[2] A tier 1 recruit is one who is an HSDG, has an adult-education diploma, or completed at least one semester of college, or attended virtual or distance learning or an adult or alternative school. See Office of the Under Secretary of Defense for Personnel and Readiness, 2016, Appendix A, p. 13.

of total contracts and one for the number of high-quality contracts (Appendix D).

What Is the Right Level of Recruit Quality, and Is Regular Military Compensation a Cost-Effective Way to Achieve That Quality?

Rigorous studies have found that higher-quality personnel perform better in the military. But it is hard to know how much quality is optimal (i.e., the right balance between the gain in defense capability from more high-quality recruits and the cost of higher RMC and recruiting and retention resources). Perhaps changes in the defense environment have shifted requirements toward recruits with higher scores on the Armed Forces Qualification Test, and the 70th percentile might not be the right standard today. Further, past research has shown that higher-than-ECI increases in basic pay were critical to helping the services manage recruiting and retention when frequent, long deployments to Iraq and Afghanistan strained recruiting and retention (Hosek and Martorell, 2009; Asch, Heaton, et al., 2010). Research also shows that RMC is a blunt and costly instrument for addressing recruiting challenges because it is not targeted and affects the personnel budget of every service.

If RMC decreased from its current level relative to civilian pay, the Air Force, Navy, and Marine Corps might reduce recruit quality. Still, if today's recruit quality is needed for today's manning requirements, the services would need to increase recruiting resources and special and incentive pays, such as enlistment bonuses, to make up for the decrease in RMC. The Army's response would be similar. If RMC were to decrease, the Army, too, would need to compensate for this by increasing recruiting resources. But, in the Army's case, this would be necessary also to prevent recruit quality from falling below the DoD quality guidance.

Acknowledgments

We are pleased to thank the Office of the Under Secretary of Defense for Personnel and Readiness Office of Compensation for sponsoring this report. We especially appreciate the guidance offered by Jeri Busch, director for military compensation, and Don Svendsen of the Office of Compensation, as well as Don's comments. We are grateful to Mike DiNicolantonio and his team at the Research, Surveys, and Statistics Center of the Office of People Analytics in the Defense Human Resources Activity for tabulations on educational attainment of those in the military. At RAND, David Knapp helped develop the database for this project. He and Bruce R. Orvis provided valuable comments on an earlier draft of this report, and Dave later served as a reviewer of the report. Dave's and John Warner's (also of RAND) reviews of an earlier version of this report were extremely helpful in pointing the way to valuable improvements. Christine DeMartini and Craig Martin helped process the military pay and Current Population Survey files, and Lisa Bernard edited the report.

Abbreviations

AFQT	Armed Forces Qualification Test
ARMS	Assessment of Recruit Motivation and Strength
ASVAB	Armed Services Vocational Aptitude Battery
BAH	basic allowance for housing
BAQ	basic allowance for quarters
BAS	basic allowance for subsistence
CPS	Current Population Survey
DEP	delayed-entry program
DoD	U.S. Department of Defense
ECI	Employment Cost Index
FY	fiscal year
HSDG	high school diploma graduate
MAC	military annual compensation
MEPS	military enlistment processing station
NCES	National Center for Education Statistics
NDAA	National Defense Authorization Act
NLSY	National Longitudinal Survey of Youth

NPS	non–prior service
NR	not reported
QRMC	Quadrennial Review of Military Compensation
RMC	regular military compensation
TAPAS	Tailored Adaptive Personality Assessment System
TTAS	Tier 2 Attrition Screening
VHA	variable housing allowance
YOS	year of service

CHAPTER ONE

Introduction

Military pay is a policy tool vital for ensuring the success of the all-volunteer force. Researchers have consistently found that the enlistment of high-quality recruits and their retention in the military are responsive to increases in military pay.[1] Given the pivotal role of military pay, policymakers face the ongoing question of whether military pay is adequate.

Regular military compensation (RMC) is a useful measure of military pay. RMC includes basic pay, basic allowance for housing (BAH), basic allowance for subsistence (BAS), and the federal tax advantage of the allowances, which are tax free. RMC accounts for approximately 90 percent of current cash compensation (Office of the Under Secretary of Defense for Personnel and Readiness, 2012a, Chapter 2, p. 17).[2] Basic pay and BAH are RMC's largest components, while BAS is only a small percentage of it.

---

[1] Asch, Hosek, and Warner, 2007, surveys the literature and reports estimates of recruiting and retention responsiveness to military pay. More-recent estimates are in Asch, Heaton, et al., 2010.

Military recruits are deemed high quality if they are high school diploma graduates (HSDGs) and score in the upper half of the Armed Forces Qualification Test (AFQT) score distribution.

[2] The elements of current compensation in RMC are apart from expenditures for Social Security tax, which the U.S. Department of Defense (DoD) pays on behalf of the service member; the charges and contributions toward military retirement and health benefits; overseas housing allowance; uniform allowance; special and incentive pays; and separation pay. Active-component members not living in government-furnished housing receive basic pay, BAS, and BAH. In pay comparisons, the value of government-furnished housing is imputed as the value of BAH. See Office of the Under Secretary of Defense (Comptroller), 2017.

The Ninth Quadrennial Review of Military Compensation (QRMC) (Office of the Under Secretary of Defense for Personnel and Readiness, 2002) advised,

> Military and civilian pay comparability is critical to the success of the All-Volunteer Force. Military pay must be set at a level that takes into account the special demands associated with military life and should be set above average pay in the private sector. Pay at around the 70th percentile of *comparably educated civilians* [italics added] has been necessary to enable the military to recruit and retain the quantity and quality of personnel it requires. (Office of the Under Secretary of Defense for Personnel and Readiness, 2002)

In the 9th QRMC's assessment, trends in education meant that military pay was not keeping pace with private-sector compensation for midgrade enlisted members and junior officers. The 9th QRMC provided data showing that, although most enlisted personnel entered service with a high school education, about half of midcareer enlisted (E-5, E-6, and E-7) had attained "some college," and, among the most-senior enlisted (E-8 and E-9), about 50 percent had some college and 25 percent had four or more years of college. Enlisted RMC was at about the 70th percentile of wages for full-time, full-year male workers with high school educations but, for enlisted with some college and with ten to 20 years of experience, RMC was at only the 50th percentile of wages for full-time, full-year male workers with some college. These findings were for 1999, a boom year when real median household income was at its peak, having risen steadily since 1993 (Federal Reserve Bank of St. Louis, undated [b]). Although this context could be expected to lead to lower RMC percentiles, the 9th QRMC's basic points were independent of economic conditions: Enlisted personnel were adding to education throughout their careers, and, given the many enlisted with some college, RMC should be at about the 70th percentile of full-time, full-year civilian workers with some college. The 9th QRMC made a similar argument for officers, many of whom were obtaining education beyond bachelor's degrees.[3]

The 9th QRMC's report supported several actions mandated by the National Defense Authorization Act (NDAA) for Fiscal Year (FY) 2000. These were a higher-than-usual 4.8-percent increase in basic pay for FY 2000; a structural adjustment to the basic-pay table, with targeted pay raises in grades E-5 through E-7 in July 2001; increases in basic pay through FY 2006 that would be 0.5 percent above private-sector pay increases; and increases in the amount of BAH ("2001 US Military Basic Pay Charts," undated; Office of the Under Secretary of Defense for Personnel and Readiness, undated).[4]

A decade later, the 11th QRMC found that, in 2009, RMC was at "about the 90th percentile of equivalent civilian wages" for the combined civilian comparison groups it chose for enlisted personnel comparisons. These groups were civilians with high school diplomas, those with some college, and those with associate's degrees. For officers, the comparison groups were those with four-year college degrees and those with master's degrees or higher, and RMC was "at about the 83rd percentile" for these groups combined (Office of the Under Secretary of Defense for Personnel and Readiness, 2012a).[5]

[3] The 9th QRMC compared RMC to the civilian wages of men, while the 11th QRMC used a weighted average of the wages of men and women, with the weights reflecting the gender mix in the military. In this report, we use a gender weighting. However, we also made pay comparisons by gender and found that the difference between RMC percentiles for mixed gender was only slightly higher than that for men only.

[4] The BAH increases decreased the expected out-of-pocket costs for housing from 20 percent in 2000 to zero in 2005. However, recent policy changes are increasing the out-of-pocket expense for housing. It "will increase by one percent annually until it is capped at 5%. Thus, out-of-pocket expenses are 2% in 2016, 3% in 2017, 4% in 2018 and 5% in 2019" (Defense Travel Management Office, 2018, p. 7).

[5] The 11th QRMC also recognized the growing prevalence of higher education among senior enlisted personnel, as the 9th QRMC had noted. Thus, the 11th review compared RMC for senior enlisted personnel to pay for civilians with two- and four-year degrees. This comparison showed that RMC over years of service (YOSs) 15 through 23 was about $20,000 higher in 2009 dollars than the median earnings of civilians with four-year degrees, and it rose to $40,000 higher at YOS 30. This increase was thought to reflect promotion to higher pay grades among those remaining in the military. The 11th review did not report a percentile for this comparison.

The 10th QRMC recommended a more comprehensive measure of military pay called military annual compensation (MAC). MAC includes RMC, as well as state and Federal Insurance Contributions Act tax advantage (26 U.S.C. Ch. 21), the benefits of avoiding the cost of health care, and the value of the military retirement benefit. The tenth review observed that MAC should be comparable to the 80th percentile of civilian earnings (DoD, 2008). DoD opted to continue to use RMC rather than MAC, however. According to the U.S. Government Accountability Office, this was because DoD

> views its compensation as directly related to its ability to meet recruiting and retention goals. As a result, the department would rather rely on a known measure—regular military compensation compared to cash compensation for civilians—than to base its comparisons on a measure that is unknown and could vary depending on methodology used to estimate the value of benefits. . . . [C]omparing civilian cash compensation with regular military compensation allows for a more homogeneous comparison of military and civilian compensation. (U.S. Government Accountability Office, 2010, p. 19)

The use of civilian pay to benchmark military pay seems to have its origins in the 1948 Advisory Commission on Service Pay (commonly known as the Hook Commission). Its report established that military compensation rates should be based on comparisons between military and private-sector pay among those with similar levels of responsibility. But pay comparison is not an end in itself. As later commissions and studies have consistently argued, pay is a tool for meeting manpower requirements and, as stated by the Congressional Budget Office, "the best barometer of the effectiveness of DoD's compensation system may be how well the military attracts and retains high-quality personnel" (Murray, 2010, p. 1). The level of pay required to meet military requirements might or might not be at a particular percentile. Although they embrace this point, QRMCs remain interested in how well military pay compares to civilian pay, as seen in comparisons done for the 9th, 10th, and 11th QRMCs.

The research summarized in this report focused on how military and civilian pay compare and how military and civilian pay relate to recruiting outcomes. The questions we address are as follows:

- How does military pay in 2016 compare to civilian pay, and, in particular, is military pay above the 70th percentile of civilian pay?
- How do the results for 2016 compare to those for 2009, when the 11th QRMC measured percentiles?
- Insofar as military pay has increased more than civilian pay has since 1999, when the 9th QRMC measured percentiles, to what extent is this increase associated with an increase in the quality of recruits?

We followed the approach taken by the 9th and 11th QRMCs and measured military pay by RMC. We compared RMC in 2016 for enlisted and officers, by YOS, to civilian pay for different civilian comparison groups. In a departure from the 11th review, we used more-detailed information on educational attainment for enlisted over the career. We also obtained similar information for 2009, permitting an apples-to-apples comparison *by YOS* for 2016 and 2009.

We also computed the RMC percentile *over time* for the civilian comparison groups used in the 11th QRMC's over-time comparisons—namely, workers with high school, some college, and associate's degrees. These comparisons allowed us to focus on specific education groups and to examine trends for each service. We further compared RMC to the Employment Cost Index (ECI) and discuss the limitations of such comparisons (Appendix C).

To address the third question, we estimated regression models, by service, of the relationship between measures of recruit quality and the RMC/wage ratio, controlling for other factors, including recruiting goal, deployment, unemployment, and gender. We considered recruit quality for two reasons. First, analyses (e.g., Winkler, Fernandez, and Polich, 1992; Orvis, Childress, and Polich, 1992; "Project A," 1992; Scribner et al., 1986) have shown that performance on mission-essential tasks increases with AFQT, which is a factor in recruit quality.

Also, high-quality recruits are more likely than other recruits to complete their enlistment contracts (e.g., Buddin, 1984, 2005).[6] Second, in terms of AFQT, the quality a service brings in is approximately the quality it retains throughout the enlisted career (Asch, Romley, and Totten, 2005). Thus, recruit quality also provides a metric, although not the only metric, of the quality of the overall enlisted force. As a way of illustrating the results, we used the estimated models to predict the change in recruiting outcomes by AFQT category given the RMC/wage ratio in 1999 and 2015, the lowest and highest years of the ratio, while holding other variables at given values.

Chapter Two compares RMC to the pay of civilians with similar characteristics. Chapter Three focuses on whether RMC increases since 1999 are associated with an increase in the quality of recruits. Chapter Four summarizes and discusses the main findings.

---

[6] Attrition is approximately 1.5 to two times higher for non–high school diploma graduates (HSDGs) than for HSDGs. Janice H. Laurence found attrition rates at the end of three YOSs to be 22 percent for HSDGs, 45 percent for high school–equivalent earners, and 43 percent for non-HSDGs (Laurence, 1984). Richard Buddin, studying attrition in the first six months, found rates roughly twice as high for non-HSDGs as for HSDGs (Buddin, 1984). In a later study, he found six-month Army attrition rates for the FY 1995–FY 2001 cohorts to be about 20 percent for high school–equivalent earners, 14 percent for HSDGs, and 12 percent for seniors and 36-month attrition rates to be about 50 percent for high school–equivalent earners, 34 percent for HSDGs, and 31 percent for seniors (Buddin, 2005).

CHAPTER TWO

Military and Civilian Pay Comparisons and Regular Military Compensation Percentiles

This chapter presents comparisons of military and civilian pay. We discuss our data sources and present comparisons of RMC to civilian wages over a career, controlling for education. We compare these results to those of the 11th QRMC. We then show trends over time in military pay compared to civilian pay, by service and for specific age and education groups.

Data Sources

The measure of military pay that we used is RMC. We obtained estimates of RMC from two sources. For comparisons over a career, we took RMC from the DoD Directorate of Compensation's *Selected Military Compensation Tables* (Greenbook) (Directorate of Compensation, 2017). In it, RMC is an average across pay grade and dependent status at each YOS (Directorate of Compensation, 2017). For trend comparisons, we used the median weekly pay for specific service/age/education/gender groups computed using the Active Duty Pay Files provided by the Defense Manpower Data Center.[1]

[1] The Greenbook RMCs are averages, and the RMCs from the Active Duty Pay Files are medians. This should make little difference: "Military wages are not skewed, because no service members receive an inordinately high wage based on RMC. As a result, the mean and median are roughly the same" (Office of the Under Secretary of Defense for Personnel and Readiness, 2012a).

Computing RMC with the military-pay files required that we compute the relevant tax advantage.[2] It is based on taxable (basic pay) and nontaxable (BAS and BAH) income, number of dependents, and marital status.[3]

A key characteristic in comparing military and civilian pay is education. We used the education distribution of officers and enlisted personnel from the August 2009 and September 2016 Status of Forces Surveys of Active Duty Members, provided by the DoD Office of People Analytics.

We measured civilian pay as weekly pay and compared military personnel to civilian workers with similar characteristics. Data on wages and characteristics are from the Current Population Survey (CPS) Annual Social and Economic Supplement, also known as the

---

[2] The tax advantage can be thought of as the amount of additional basic pay that would have to be paid to a member to hold that member harmless if BAS and BAH were taxable income. We used a simple line-search algorithm to solve for the numerical value of tax advantage. The federal income tax calculations include the cutoff ranges for each tax rate. If a person moved from one bracket to the next when we included the tax advantage, we calculated the portion of income plus tax advantage that fell under the original bracket at that rate and the remaining income for the next bracket. As income increases, we also adjusted payments through FICA and the Earned Income Tax Credit (EITC).

We included data for people with less than 12 months of service in a given year, which could have biased our pay estimates downward. We based the calculations on monthly pay files—one record per person per month. We annualized dollar amounts by multiplying pay by 12 and dividing by the number of months for which data were present in a given year. It is possible that including people with less than 12 months of service, like we did, can deflate total pay even with the annualization. This can be caused by the fact that people who leave during the year are less likely to be present in the year after promotion (and might even be less likely to be promoted).

[3] Negative pay amounts in the pay files represent a take-back from an earlier overpayment. We summed them as is to roll up to an annual level, then divided to get a weekly average. For those without reliable location information (e.g., those overseas, people on ships), we imputed BAH from the BAHs of those in the same grade and with the same dependent status averaged over all locations. For members without BAH entries (those who live in on-base housing), we imputed BAH from those in the same grade, dependent status, and location (ZIP code) cells. BAH replaced basic allowance for quarters (BAQ) and variable housing allowance (VHA) in 1998; however, the old BAQ and VHA fields still exist in the pay files through 2012 and contain values, while BAH is rarely populated. Thus, in computing RMC, we summed BAH, BAQ, and VHA to get a proper amount prior to 2013. We did not take the overseas housing allowance into account.

March CPS. The CPS uses a representative random sample of the population and is administered by the Bureau of Labor Statistics.

Control Characteristics in Military and Civilian Pay Comparisons

As the 9th and 11th QRMCs note, key controls for comparing RMC and civilian pay are labor-force participation, gender, education level, and years of labor-force experience.[4] It is also important to consider that the CPS top-codes high wages.[5]

Labor-Force Participation and Gender

Like the 11th QRMC, we used data on full-time, full-year workers. A full-time, full-year worker is one with a usual work week of more than 35 hours and who worked more than 35 weeks in the year.[6] Also, like the 11th QRMC, we weighted civilian-wage data by the percentages of men and women in the military. In 2015, the percentages were 85 percent men and 15 percent women for enlisted and 83 percent men and 17 percent women for officers (Office of the Deputy Assistant Secretary of Defense for Military Community and Family Policy, 2016, pp. 18–19).

---

[4] The pay comparisons done for past QRMCs have taken a national perspective, which is relevant for an overview and responsive to the fact that service members are periodically reassigned to different locations, suggesting the importance of assessing how military pay compares, on average, to pay in the economy. There might also be geographical differences in wages, and the military adjusts for these to some extent through housing allowances, which are locality specific.

[5] Annual wages in the CPS are reported up to a limit, the top code. Wages above the top code are censored, meaning that they are set equal to the top code.

[6] Changing the criterion to a usual work week of 40 or more hours gives similar results because few workers report usual work weeks of 36 to 39 hours. Changing weeks worked to 48 or more also gives similar results, given the criterion of a usual work week of more than 35 hours.

Education Level

We used five education levels for enlisted personnel: high school, some college (more than high school but no degree), associate's, bachelor's, and master's or higher.[7] We used two levels for officers: bachelor's and master's or higher. The 11th QRMC used three education levels for examining enlisted pay trends over time: high school, some college, and associate's degree. These comparison groups underlie the 11th QRMC's finding that RMC for enlisted was at about the 90th percentile of pay for comparable civilian groups.[8] To enable comparison with the 11th QRMC, we also computed the RMC percentile using the three groups it used.

Enlisted

Table 2.1 shows educational attainment for enlisted members for 2009 and 2016. The row entries in the table sum to 100 percent with rounding.

In 2016, most enlisted personnel reported entering the military with a high school education or some college.[9] The percentage with high school decreases with rank as enlisted add to their education in service. The percentage with some college, including both those with less than one year of college and those with more than one year of college but no degree, at first increases with rank then decreases, while the percentage with an associate's degree or more increases steadily

---

[7] Because the question on the CPS asks about the completed level of education, people who have completed coursework beyond bachelor's degrees but have not yet attained master's or professional degrees are included with those who have bachelor's degrees.

[8] The 11th review also presented a chart comparing RMC for senior enlisted to the median wage of workers with four-year college degrees, but workers with four-year and advanced degrees were not included in the computation that led to the finding that enlisted RMC was at the 90th percentile.

[9] In the Status of Forces Survey, *high school education* includes equivalency (e.g., GED) certificates. However, in categorizing recruits, the services distinguish between tier 1 and tier 2:

> Tier 1 accessions are primarily HSDGs, but they also include people with educational backgrounds beyond high school, as well as those with adult education diplomas, one semester of college, and, in recent years, some home-schooled and virtual/distance learning graduates. (Office of the Under Secretary of Defense for Personnel and Readiness, 2016, Table D.7)

Tier 2 includes equivalencies.

Table 2.1
Educational Attainment of Enlisted Personnel, by Pay Grade, 2009 and 2016, as Percentages

Pay Grade	Non–High School Graduate		High School Graduate		Less Than One Year of College		One or More Years of College, No Degree		Associate's Degree		Bachelor's Degree		Master's Degree or Higher	
	2009	2016	2009	2016	2009	2016	2009	2016	2009	2016	2009	2016	2009	2016
E-2	1	NR	70	72	20	16	8	9	NR	1	1	1	NR	NR
E-3	1	0	48	49	23	21	21	18	4	7	3	4	0	0
E-4	0	1	39	33	25	24	22	26	7	8	6	8	1	1
E-5	1	1	25	22	22	16	32	35	13	17	6	8	0	1
E-6	1	1	17	14	23	13	30	30	20	26	8	14	1	2
E-7	1	0	10	7	15	10	30	26	28	32	14	20	2	6
E-8	0	0	9	4	13	6	30	26	24	26	20	27	4	11
E-9	NR	0	7	6	10	4	17	12	22	25	30	33	14	20

SOURCES: Office of People Analytics, 2009; Office of People Analytics, 2016. The tabulation is based on the 2016 Status of Forces Survey.

NOTE: NR = not reported. The percentages in each row sum to 100 with rounding. There is no row for E-1s because their education distribution was not reported in the survey. In this table, *high school graduate* includes traditional diploma and alternative diploma (e.g., home school, equivalency test, distance learning). The survey responses are weighted to be representative of the force.

with rank. The percentage with an associate's degree stabilizes at 25 to 30 percent for E-6 and higher ranks, and the percentage with a bachelor's or more climbs from 16 percent at E-6 to 53 percent at E-9.

Table 2.1 shows a similar pattern for 2009. However, the percentages of enlisted at higher levels of education were higher in 2016 than in 2009. In 2016, 20 percent of E-7s reported having bachelor's degrees, versus 14 percent in 2009, and 20 percent of E-9s reported having advanced degrees—master's, PhD, or professional degrees—versus 14 percent in 2009.

The 9th QRMC identified the trend toward obtaining more education in service. With many enlisted members adding to their education in the military, the team reasoned, education beyond high school was increasingly relevant to judge the adequacy of military pay. Data in the 9th QRMC report are more limited, but we could compare the percentage of enlisted in two education categories: those with some college or associate's degrees and those with bachelor's degrees or higher.[10] The results are in Table 2.2.

Reported levels of education were considerably higher in 2016 than in 1999 or 2009. The percentage of E-4s with some college or associate's degrees grew from 31 percent in 1999 to 54 percent in 2009 and 58 percent in 2016. The percentage with bachelor's or higher grew from 5 percent in 1999 to 7 percent in 2009 and 9 percent in 2016. The percentage of E-6s with some college or associate's degrees grew from 57 percent in 1999 to 73 percent in 2009 but was 69 percent in 2016. The percentage of E-6s with bachelor's degrees or higher changed from 10 percent in 1999 to 9 percent in 2009 (not a statistically significant decline), then rose to 16 percent in 2016. The decline in 2016 (from 2009 levels) in associate's degrees for E-6 or more might reflect a substitution toward bachelor's or higher education. E-7s, E-8s, and E-9s were also much more likely to have bachelor's or more in 2016 than in 2009 or 1999.

[10] We inferred estimates for 1999 from Figure 2-4 in the 9th QRMC report (Office of the Under Secretary of Defense for Personnel and Readiness, 2002), which is based on data from the 1999 Survey of Active Duty Personnel.

Table 2.2
Enlisted Personnel with Post–High School Education, by Pay Grade, 1999, 2009, and 2016, as Percentages

	Some College or Associate's Degree			Bachelor's Degree or Higher		
Pay Grade	1999	2009	2016	1999	2009	2016
E-1	7	NR	NR	1	NR	NR
E-2	18	28	26	0	1	1
E-3	22	48	46	2	3	4
E-4	31	54	58	5	7	9
E-5	47	67	68	6	6	9
E-6	57	73	69	10	9	16
E-7	60	73	68	18	16	26
E-8	56	67	58	22	24	38
E-9	57	49	41	27	44	53

SOURCES: Office of the Under Secretary of Defense for Personnel and Readiness, 2002, Figure 2-4; Office of People Analytics, 2009; Office of People Analytics, 2016. Data are from the Status of Forces Surveys.

NOTE: *High school graduate* includes traditional diploma and alternative diploma (e.g., home school, equivalency test, distance learning). The survey responses are weighted to be representative of the force. The 9th QRMC report presents the combined percentage of enlisted with bachelor's degrees or higher; it does not present the percentage with bachelor's only. For 2009 and 2016, Table 2.1 shows separate percentages for bachelor's and master's or higher, and this table adds those percentages to obtain bachelor's degrees or higher.

Officers

Many officers add to their education while in service.[11] Table 2.3 shows the percentages reporting the highest degrees as college degrees or advanced degrees, for 1999, 2009, and 2016 (Office of the Under Secretary of Defense for Personnel and Readiness, 2002, Figure 2-15). *College degree* includes bachelor's and associate's degrees. *Advanced degree* includes master's, doctoral, and professional school degrees.

Like for enlisted, the percentage with advanced degrees increases with rank in 1999, 2009, and 2016, and the extent of increase has trended upward. In 1999, 69 percent of O-4s had advanced degrees, versus 78 percent in 2016, for instance.

Years of Labor-Force Experience

The Greenbook provides RMC by YOS. To compare civilian pay to RMC, we needed a comparable measure of years of labor-force experience. The March CPS does not record years of labor-force experience, however, so we used assumptions to map age and years of education to years of labor-force experience. We list these in Table 2.4.

These assumptions will result in an overstatement of experience for those who choose to enroll in school at a later starting age or who enroll but attend school part time or drop out and reenroll later. That said, in 2014, 89 percent of students when first enrolled at two-year institutions were 19 years old or younger, and 85 percent of

---

[11] The services offer professional military education courses and programs, encourage officers (and enlisted) to obtain additional education from accredited institutions, and consider education in promotion decisions. The services' institutions include the Air Force Institute of Technology, the Air University, the Joint Forces Staff College, the Marine Corps University, the National Defense University, the National War College, the Naval Postgraduate School, the U.S. Army War College, and U.S. Naval War College. These institutions offer certificates of course completion, master's degrees, and, at some, PhDs. For instance, the U.S. Naval War College has programs that offer master's degrees, as well as diplomas certifying the completion of a course (U.S. Naval War College, undated). The Naval Postgraduate School has master's and PhD programs (Naval Postgraduate School, undated). The Marine Corps University includes master's programs (Marine Corps University Foundation, undated). The Army War College offers professional military education courses and has master's programs (U.S. Army War College, undated), as does the Air Force Institute of Technology (Air Force Institute of Technology, 2018). Therefore, officers have opportunity, encouragement, and incentive to obtain additional education.

Table 2.3
Educational Attainment of Officer Personnel, by Pay Grade, 1999, 2009, and 2016, as Percentages

	College Degree			Master's Degree or Higher		
Pay Grade	1999	2009	2016	1999	2009	2016
O-1	97	93	90	3	6	9
O-2	91	87	84	9	11	15
O-3	59	60	55	39	39	44
O-4	31	30	21	69	69	78
O-5	15	13	6	85	85	94
O-6	8	4	2	92	96	98

SOURCES: Office of the Under Secretary of Defense for Personnel and Readiness, 2002, Figure 2-14; Office of People Analytics, 2009; Office of People Analytics, 2016.

NOTE: *College graduate* includes bachelor's and associate's degrees. *Advanced degree* includes master's, doctoral, and professional school degrees.

Table 2.4
Assumptions About Civilian Labor-Force Experience

For Labor-Force Experience for Someone at This Level of Education Attainment	Subtract This Number from the Person's Age in Years
High school graduate	18
Some college	20
Associate's degree	20
College graduate	22
Advanced degree	24

students when first enrolled at four-year institutions were 19 years old or younger (National Center for Education Statistics [NCES], 2017). These percentages suggest that our assumptions are fairly accurate for students entering two- and four-year institutions and completing their programs within two or four years. But again, students might enroll but not complete these programs. Using NCES data, we calculated that, of high school completers enrolled in two- or four-year colleges in October of the year they completed high school, by the ages of 25 to 29, about 42 percent had completed two-year degrees and 82 percent had completed four-year degrees (NCES, 2015, 2016).[12]

The 11th QRMC defined years of labor-force experience as age minus education minus 7.[13] Neither this approach nor our approach is perfect. Under this approach, a person who graduates from high school at age 18 or college at age 22 begins with age-minus-one years of experience, which, compared with our approach, displaces the civilian wage–experience curve to the right by one year. This curve thus lies below our wage–experience curve; as a result, the RMC percentile is higher than under our approach. The difference depends on how fast the civilian wage increases with experience.

From past research, we have information on how fast the civilian wage increases with experience. For full-time, full-year workers with bachelor's degrees or higher, we have estimated elsewhere a civilian-wage increase of 5.3 percent per year at ages 25 to 29, 5.2 percent per year at ages 30 to 34, and 3.3 percent at ages 35 to 39 (Knapp, Asch, et al., 2016, p. 67). This implies that a one-year difference in experience translates to a roughly 5-percent difference in wage in the first ten or so years of experience and 3 percent in the next five years. Yona Rubenstein and Yoram Weiss found average wage growth in the first ten years of labor-force experience of 6.3 percent for college graduates, 7.7 percent for those with master's degrees or more, and 5.6 percent for

[12] The range of ages is the range given in the NCES data table.

[13] The 11th QRMC states,

> Since job experience begins at different ages for civilians, depending on their level of education, we use the civilian age, minus the normative number of years of education for whatever degree they have, minus 7 (the oldest year most children are in the first grade) as the proxy for civilian workforce experience. (Office of the Under Secretary of Defense for Personnel and Readiness, 2012b, p. 9)

By using the highest year in which most children are in first grade, this approach gives a conservative estimate of years of labor-force experience, while our approach provides a less conservative estimate. NCES does not provide data on the age distribution of high school graduates but uses age groups (e.g., showing the high school completion rate for ages 18 to 19). For example, see Chapman et al., 2011.

high school graduates (Rubenstein and Weiss, 2007). In the next five-year range (11 to 15 years of experience), the growth rates are, respectively, 5.3 percent, 4.5 percent, and 3.3 percent. These estimates differ somewhat from ours, but the impact is qualitatively similar.

Wage Top Coding

Annual wages in the CPS are reported up to a limit, the top code. Wages above the top code are censored, meaning that they are set equal to the top code. The top code varies by state, from $150,000 to $375,000, and the average for all states is $210,000. For education up to associate's degree, the top code is sufficiently high not to pose an issue when computing RMC wage percentiles. For some workers, especially college educated, civilian earnings exceed the top code. When that occurs, the CPS reports the top-coded wage instead of the actual wage. Uncorrected, the effect is to compress the wage percentile curves (e.g., the wages at the 95th percentile would appear closer to those at the 90th percentile than they actually are). This does not prove to be a problem for our enlisted or officer RMC percentile calculations, however, because RMC is well below the top code and the wage percentiles *below* the top code are not affected by top-coding.

Regular Military Compensation Percentiles, by Year of Service, 2016

Figures 2.1 and 2.2 compare enlisted and officer RMC in 2016 to civilian wages at the 50th (median) and 70th percentiles, for *given* levels of education.[14] The figures are illustrative, and the civilian-wage

[14] As mentioned, for these comparisons, we took RMC from Table B.4 in *Selected Military Compensation Tables* from 2016 (Directorate of Compensation, 2016), commonly known as the Greenbook. That table provides average RMC, by YOS, for enlisted and for officer personnel. Average RMC at a given YOS includes personnel at all pay grades at that year, so it is a comprehensive measure of average RMC by YOS. It also includes both single members and those with dependents, so it accounts for the fact that BAH differs with dependency status and that the percentage of members with dependents increases as members marry during their military careers. The computation of RMC in the first few YOSs might be affected by "partial BAH" paid to members who live on base. If no BAH was paid, we imputed full

Figure 2.1
Enlisted Regular Military Compensation, Civilian Wages, and Regular Military Compensation Percentiles for Full-Time, Full-Year Workers with High School, Some College, or Bachelor's Degrees, 2016

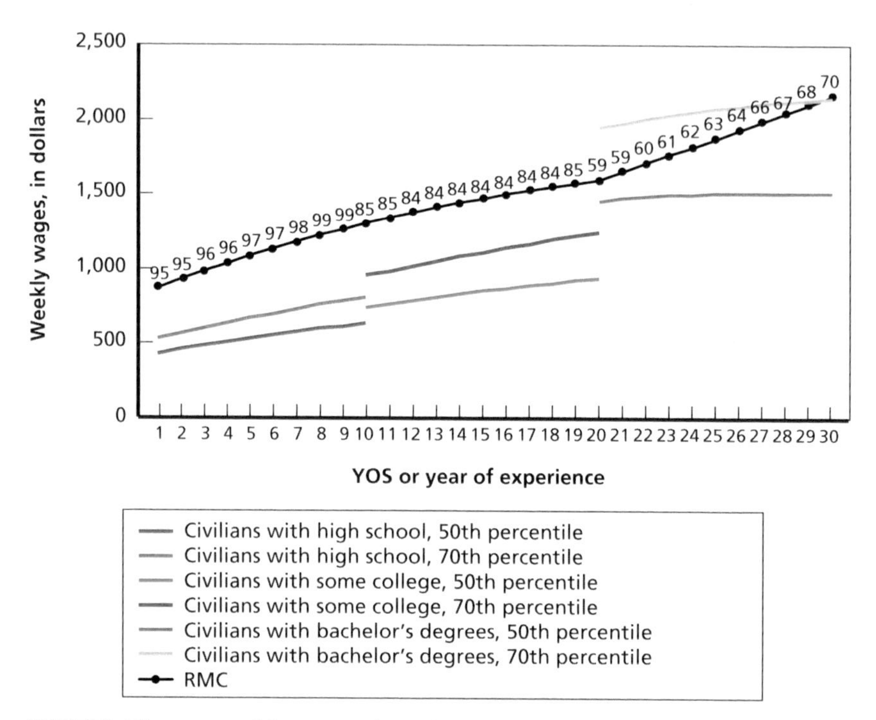

SOURCES: Directorate of Compensation, 2016; U.S. Census Bureau, 2018.
NOTE: RMC percentile varies by YOS (1–9 = high school, 10–19 = some college, and 20–30 = a bachelor's degree or higher. We weighted civilian-wage data by enlisted military gender mix. Colored lines are smoothed wage curves for the 50th and 70th percentiles of the given level of education. The black line is enlisted RMC, and the number above the black line is the percentile in the wage distribution for high school (YOSs 0 through 9), some college (YOSs 10 through 19), and bachelor's (YOSs 20 through 30).

BAH, but, if partial BAH was paid, we used it in the RMC computation. Partial BAH is substantially lower than full BAH, so we might have understated the value of housing to junior members. If so, our estimates of RMC percentile might be understated for junior enlisted personnel. As we show, RMC percentiles are relatively high for junior personnel even with the use of partial BAH.

Figure 2.2
Officer Regular Military Compensation, Civilian Wages, and Regular Military Compensation Percentiles for Full-Time, Full-Year Workers with Bachelor's Degrees or with Master's Degrees or Higher, 2016

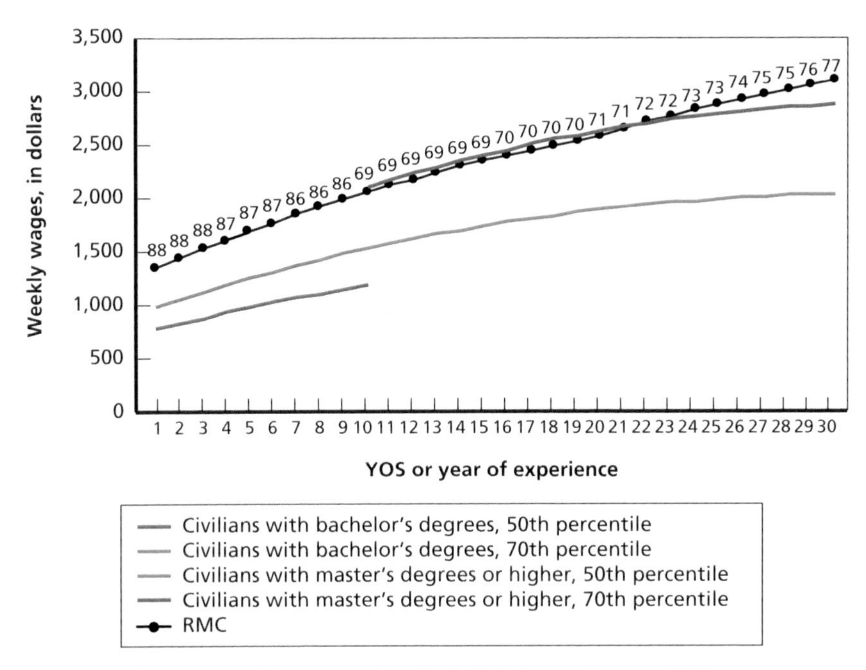

SOURCES: Directorate of Compensation, 2016; U.S. Census Bureau, 2018.
NOTE: RMC percentile varies by YOS (1–9 = bachelor's, 10–30 = master's degree or higher). We weighted civilian-wage data by military gender mix. Colored lines are smoothed wage curves for the 50th and 70th percentiles of the given level of education. The black line is enlisted RMC, and the numbers above the black line are the percentile in the wage distribution for bachelor's (YOSs 1 through 9) and master's degree or higher (YOSs 10 through 30).

lines have been smoothed with quadratic regressions on the raw data. In Tables 2.5 and 2.6, we show the RMC percentile for *each* level of education at each YOS.

For enlisted personnel, RMC is compared to the pay of workers with high school over the first nine YOSs, of workers with some college in YOSs 10 to 19, and of workers with bachelor's degrees for YOSs 20 to 30. RMC is over the 90th percentile in the first nine years and at the 84th or 85th percentile for YOSs 10 to 19. RMC is at the 59th percen-

Table 2.5
Regular Military Compensation as a Percentile of Civilian Wages, by Level of Education and Year of Service, for Enlisted Personnel, 2016

	Predicted Education Distribution, by YOS					RMC Percentile						
YOS	High School	Some College	Associate's	Bachelor's	Master's Plus	High School	Some College	Associate's	Bachelor's	Master's Plus	Weighted Average	Enlisted Count
1	0.54	0.37	0.05	0.04	0.00	88	89	81	53	19	86.5	163,380
2	0.42	0.44	0.08	0.06	0.00	94	93	96	63	32	91.6	131,827
3	0.33	0.49	0.10	0.07	0.01	95	83	92	54	33	85.5	120,633
4	0.27	0.51	0.13	0.08	0.01	98	91	84	64	47	89.4	145,655
5	0.24	0.51	0.15	0.09	0.01	93	81	78	59	33	80.9	98,591
6	0.21	0.50	0.17	0.10	0.01	92	81	92	60	36	82.6	68,394
7	0.20	0.49	0.19	0.10	0.02	94	86	90	61	46	85.1	58,219
8	0.19	0.48	0.21	0.11	0.02	92	95	86	61	40	87.8	48,227
9	0.18	0.46	0.22	0.12	0.02	91	85	79	65	35	81.4	40,824
10	0.17	0.45	0.24	0.13	0.02	90	84	82	56	41	80.1	32,903
11	0.16	0.44	0.25	0.13	0.02	91	84	76	53	35	77.8	32,326
12	0.14	0.43	0.26	0.14	0.03	92	86	80	62	37	80.6	25,801
13	0.13	0.42	0.27	0.15	0.03	91	88	80	53	35	79.4	27,428

Table 2.5—Continued

	Predicted Education Distribution, by YOS					RMC Percentile						
YOS	High School	Some College	Associate's	Bachelor's	Master's Plus	High School	Some College	Associate's	Bachelor's	Master's Plus	Weighted Average	Enlisted Count
14	0.12	0.41	0.28	0.16	0.03	90	86	81	59	37	79.1	25,937
15	0.11	0.40	0.28	0.17	0.04	95	88	78	59	43	79.2	24,976
16	0.10	0.39	0.29	0.18	0.05	94	81	75	59	35	74.5	23,044
17	0.09	0.38	0.29	0.19	0.05	89	80	81	48	43	73.1	22,176
18	0.08	0.37	0.29	0.20	0.06	91	84	83	51	36	74.8	20,109
19	0.07	0.36	0.29	0.21	0.07	90	76	86	51	35	71.9	19,696
20	0.07	0.35	0.29	0.22	0.08	94	84	85	53	40	74.8	19,425
21	0.07	0.33	0.28	0.23	0.09	90	87	88	55	40	76.0	19,767
22	0.07	0.32	0.28	0.24	0.10	94	87	91	66	46	79.5	10,228
23	0.07	0.30	0.27	0.25	0.11	91	88	91	59	36	76.0	8,448
24	0.06	0.28	0.27	0.26	0.12	93	87	93	68	50	79.5	7,385
25	0.06	0.26	0.27	0.27	0.13	96	85	82	63	51	74.3	6,103
26	0.06	0.25	0.26	0.29	0.14	94	87	93	70	48	78.5	3,696
27	0.06	0.23	0.26	0.30	0.15	95	88	90	67	41	75.5	3,506

Table 2.5—Continued

	Predicted Education Distribution, by YOS					RMC Percentile						
YOS	High School	Some College	Associate's	Bachelor's	Master's Plus	High School	Some College	Associate's	Bachelor's	Master's Plus	Weighted Average	Enlisted Count
28	0.06	0.22	0.26	0.30	0.16	95	92	83	67	53	75.9	2,537
29	0.06	0.21	0.26	0.31	0.17	96	91	90	70	46	76.9	1,960
30	0.06	0.20	0.26	0.31	0.18	96	85	87	67	55	75.4	1,578
0–20											84.2	
0–30											83.8	

SOURCES: Directorate of Compensation, 2016; DMDC data from 2016; U.S. Census Bureau, 2018.

NOTE: We computed the RMC percentile at each level of education, by YOS, as median RMC relative to the civilian wages of full-time, full-year male and female workers, weighted by their proportion in the military. We computed median RMC from active-duty pay files. Weighted average RMC percentile at each YOS is the sum of the product of the RMC percentile at a given level of education and the fraction of personnel with that level of education, shown in the left pane of the table. We estimated the education fractions using the educational attainment distribution for 2016 (see Table 2.1) and the joint distribution of personnel by pay grade and YOS from the Greenbook for 2016. The overall RMC percentiles for YOSs 0–20 and YOSs 0–30 are weighted averages of the average RMC percentile at each YOS, with weights based on the fraction of personnel count by YOS (the "Enlisted Count" column).

Table 2.6
Regular Military Compensation as a Percentile of Civilian Wages, by Level of Education and Year of Service, for Officers, 2016

	Predicted Education Distribution[a]		RMC Percentile			
YOS	Bachelor's	Master's Plus	Bachelor's	Master's Plus	Weighted Average	Officer Count
1	0.89	0.11	74	53	71.6	8,306
2	0.81	0.19	91	80	88.9	8,950
3	0.74	0.26	90	75	86.1	8,755
4	0.67	0.33	93	75	87.1	9,552
5	0.61	0.39	89	72	82.4	9,762
6	0.55	0.45	87	69	78.9	9,557
7	0.50	0.50	89	77	83.0	9,106
8	0.45	0.55	87	74	79.8	8,696
9	0.40	0.60	91	72	79.6	8,485
10	0.36	0.64	87	72	77.4	7,949
11	0.32	0.68	86	61	69.0	7,436
12	0.28	0.72	88	65	71.5	7,080
13	0.25	0.75	81	67	70.5	6,919

Table 2.6—Continued

	Predicted Education Distribution[a]		RMC Percentile			
YOS	Bachelor's	Master's Plus	Bachelor's	Master's Plus	Weighted Average	Officer Count
14	0.22	0.78	86	70	73.5	6,569
15	0.20	0.80	87	72	74.9	6,431
16	0.17	0.83	84	67	69.9	6,406
17	0.15	0.85	83	71	72.8	6,315
18	0.13	0.87	81	63	65.4	6,134
19	0.12	0.88	83	73	74.2	6,162
20	0.11	0.89	83	68	69.6	6,174
21	0.10	0.90	80	67	68.2	5,915
22	0.09	0.91	86	71	72.3	4,954
23	0.08	0.92	81	69	70.0	4,724
24	0.08	0.92	91	81	81.8	4,403
25	0.07	0.93	89	82	82.5	4,216
26	0.07	0.93	90	66	67.7	3,528
27	0.07	0.93	85	76	76.6	3,145
28	0.07	0.93	90	79	79.8	3,021

Table 2.6—Continued

	Predicted Education Distribution[a]		RMC Percentile			
YOS	Bachelor's	Master's Plus	Bachelor's	Master's Plus	Weighted Average	Officer Count
29	0.08	0.92	88	73	74.1	2,474
30	0.08	0.92	84	78	78.5	1,914
0–20					77.2	
0–30					76.7	

SOURCES: Directorate of Compensation, 2016; DMDC data from 2016; U.S. Census Bureau, 2018.

NOTE: We computed the RMC percentile at each level of education, by YOS, as median RMC relative to the civilian wages of full-time, full-year male and female workers, weighted by their proportion in the military. We computed median RMC from active-duty pay files. Weighted average RMC percentile at each YOS is the sum of the product of the RMC percentile at a given level of education and the fraction of personnel with that level of education, shown in the left pane of the table. We estimated the education fractions using the educational attainment distribution for 2016 (see Table 2.1) and the joint distribution of personnel by pay grade and YOS from the Greenbook for 2016. The overall RMC percentiles for YOSs 0–20 and YOSs 0–30 are weighted averages of the average RMC percentile at each YOS, with weights based on the fraction of personnel count by YOS (the "Officer Count" column).

[a] This is the fraction of officers, by education level, at each YOS.

tile at YOS 20 and climbs to the 70th percentile at YOS 30. (Table 2.1 uses five levels of education for enlisted, high school, some college, associate's, bachelor's, and master's or higher and does so for all YOSs.)

For officers, RMC is compared to wages of civilians with bachelor's degrees in the first nine years. In this range, RMC is at the 86th to 88th percentile of civilian pay. At ten years or more, the comparison is to wages of civilians with master's degrees or higher. RMC is at the 69th to 71st percentile from years 10 to 20 and rises to the 77th percentile by year 30. Again, this increase results in part from the selective retention of personnel promoted to higher grades. Thus, RMC for officers is initially above the 70th percentile, around it in years 10 to 25, and then above it.

Regular Military Compensation Percentiles for 2016

We computed the RMC percentiles for 2016 by YOS for each education level. This way, we could examine the RMC percentile in detail by education level. In addition, we estimated the percentage education distribution at each YOS and used this to compute the average RMC percentile by YOS, then used the number of personnel by YOS to compute an overall weighted average of the RMC percentile.[15] Tables 2.5 and 2.6 are for enlisted and officers, respectively. The RMC percentiles

[15] To compute the weighted averages, we first translated the education distribution by rank (Tables 2.1 and 2.3) to a distribution at each YOS, by interpolation. We did this in several steps. First, we obtained the joint distribution of personnel by pay grade and YOS from the Greenbook. This allowed us to compute the percentage of personnel at each pay grade, by YOSs. Second, we used these percentages to obtain a weighted average of the education distribution at each YOS (i.e., the percentage with high school, some college, associate's degrees, bachelor's degrees, and master's degrees or higher). Third, for each level of education (e.g., high school, some college), we fitted a polynomial curve to its percentages by YOS and then used the fitted curves to predict the percentage, in effect smoothing the percentages. The set of curves for the different levels of education gave us the predicted education distribution by YOS. The predicted education distribution is shown in Tables 2.5 and 2.6 for enlisted and officers, respectively. To check for sensitivity, we perturbed the education percentages by YOS and found little change in the predicted overall RMC percentile.

vary somewhat from year to year because of sample-size variation in the civilian-wage data.[16]

We found that RMC in 2016 was at about the 84th percentile for enlisted personnel with between zero and 20 YOSs, as well as for those with zero to 30 YOSs. With respect to workers with high school, RMC is around the 90th percentile; compared to wages of workers with some college, it ranges mostly between the 80th and 90th percentiles. A simple weighted average, using the enlisted count by YOS in the rightmost column, produces average percentiles of 93 for high school, 87 for some college, and 85 for associate's.[17] At a given level of education (e.g., some college), the RMC percentile is fairly stable as the experience level of military personnel increases. At the same time, the average RMC percentile—across all levels of education—decreases as experience increases, and this can be attributed to the increase in education with experience.

RMC for officers in 2016 was at about the 77th percentile for those with zero to 20 YOSs and the 77th percentile for zero to 30 YOSs. Here, a simple weighted average gives the 86th percentile for officers with bachelor's and the 70th for officers with master's or higher.[18]

We have also done the percentile calculation separately for men and women. As expected, the mixed-gender percentile is higher than the male-only percentile because civilian women's wages are lower than civilian men's wages. That said, the differences in the percentiles were not large. For instance, the average difference between the mixed-gender percentile and the male-only percentile was 0.85 percentile for enlisted HSDGs and 1.61 for enlisted with some college and 1.65 per-

---

[16] The percentiles in the tables are raw, although we have done the same calculation with smoothed percentiles and obtained overall RMC percentiles quite close to those shown.

[17] A more refined weighted average, using enlisted counts by YOS allocated according to the percentage in each education level, produces extremely similar results (e.g., average percentiles of 92 for high school, 87 for some college, and 84 for associate's).

[18] A refined weighted average for officers produces percentiles of 87 for officers with bachelor's degrees and 70 for officers with master's degrees or higher.

centiles for officers with bachelor's only and 2.64 percentiles for officers with master's degrees or higher.[19]

Comparison to Regular Military Compensation Percentiles for 2009

The 11th QRMC, using 2009 data, placed RMC at the 90th percentile of civilian pay for enlisted and the 83rd for officers. Our percentiles for 2016—the 84th for enlisted and 77th for officers—are somewhat lower than those of the 11th QRMC. Although the estimates differ, both estimates show relatively high percentiles, yet methodological differences contribute to the discrepancy.[20]

When we limited education to the categories used by the 11th QRMC, we obtained an overall RMC percentile of 88 for 2016 and 86 for 2009 for enlisted personnel, for both zero to 20 YOSs and zero to 30 YOSs. The 86th percentile is 4 percentile points less than the 90th percentile reported by the 11th QRMC for 2009. When we used the more-complete educational attainment distribution for 2009, we found that enlisted RMC was at about the 84th percentile in 2009. This is further below the 11th QRMC's 90th percentile, as might be expected, but is the same as we obtained for 2016.

There is a further point to consider, which is that Tables 2.2 and 2.3 show a change in the education distribution between 2009 and 2016. So, as a refinement, we also calculated the 2009 percentiles using the education distribution for 2016. This controls for the rise in educational attainment between 2009 and 2016. We found that enlisted

---

[19] Pay comparisons could also be extended to control for occupation (e.g., Hosek, Peterson, et al., 1992; Asch, Hosek, and Warner, 2001).

[20] These differences relate to labor-force experience and education levels. Our method for imputing labor-force experience differed from the 11th QRMC approach; we used more education levels for enlisted and included bachelor's and master's or higher; and we used the military education distribution, while the 11th QRMC applied the civilian education distribution, which likely showed less increase in education than in the military. Each of these differences contributed to our overall RMC percentile being lower than that of the 11th QRMC. (Regarding its use of the civilian education distribution, see Office of the Under Secretary of Defense for Personnel and Readiness, 2012b, p. 21, footnote 21.)

RMC in 2009 was at the 82nd percentile for zero to 20 YOSs and at the 81st percentile for zero to 30 YOSs. This is lower than our 84th percentile for 2016, for both zero to 20 YOSs and zero to 30 YOSs. Thus, when we used the more-detailed education distribution and held the education distribution constant, RMC grew a little faster than civilian pay for enlisted personnel between 2009 and 2016.

For officers, the education categories we used were the same as those used by the 11th QRMC. For 2016, we obtained the 77th percentile for YOSs 0 to 20 and the 76th percentile for YOSs 0 to 30. For 2009, our percentiles were slightly higher—namely, the 78th and the 77th, respectively. These percentiles for 2009 are lower than the 83rd percentile found by the 11th QRMC for officers. Because the education categories are the same, we ascribe the difference between our estimate and the QRMC estimate to how the number of years of labor-force experience is imputed and the 11th QRMC's use of the civilian education distribution versus our use of the military education distribution.

As with enlisted personnel, we calculated the 2009 officer percentile using the 2016 education distribution for officers to control for educational attainment between 2009 and 2016. We found officer RMC in 2009 at the 77th percentile for YOSs 0 to 20 and at the 76th percentile for YOSs 0 to 30—the same values as for 2016. Thus, when we held the education distribution constant, the RMC percentiles for officers were the same in 2009 and 2016.

In summary, when we held education constant, RMC held its overall ranking relative to civilian compensation for officers and increased slightly for enlisted from 2009 to 2016. During this period, basic pay increased by more than the ECI in FY 2010 but by less than the ECI in FYs 2014, 2015, and 2016.[21] The ECI might not be a reliable guide for military-to-civilian pay comparisons because pay change

---

[21] Basic-pay increases take into consideration the 15-month lagged ECI increase from October to October for private wage and salary workers. That is, in considering the basic-pay increase discussed in Congress in, for example, FY 2009 to take effect in January 2010, Congress used the percentage change in ECI from October 2007 to October 2008. The ECI values can be found at Federal Reserve Bank of St. Louis, undated (a). The changes by year in basic pay and lagged ECI are, respectively, 2010, 3.4 and 2.6 percent; 2011, 1.4 percent

can differ by age and education and other factors (e.g., the occupation mix can change) (discussed further in Appendix C).

Regular Military Compensation Percentile Trends for Selected Age/Education Groups, 2000–2016

We computed RMC percentile for 2000 through 2016 for education levels of high school, some college, bachelor's degree, and master's degree or higher, crossed by groups for ages 18 to 22, 23 to 27, 28 to 32, 33 to 37, and 38 to 42. We did this for all services and, for brevity, present results for Army men, although results for the other services are similar.[22] Here, we computed RMC from Defense Manpower Data Center Active Duty Pay Files.

Figures 2.3 through 2.6 show results for Army men in the following groups:

- enlisted members ages 23 to 27 who were high school graduates
- enlisted members ages 28 to 32 with some college
- officers ages 28 to 32 with bachelor's degrees
- officers ages 33 to 37 with master's degrees or higher.

These age ranges cover a high fraction of enlisted and officer personnel.[23] We have not smoothed the wage percentiles in these figures.

Age-range comparisons differ from YOS comparisons. This is because some personnel enter service at older ages, and, as a result, they have fewer YOSs than one might expect from age alone. In 2009, for example, 25 percent of enlistees in their first YOSs were over age 22

---

each; 2012, 1.6 and 1.7 percent; 2013, 1.7 and 1.6 percent; 2014, 1.0 and 1.7 percent; 2015, 1.0 and 2.1 percent; 2016, 1.3 and 2.2 percent; and 2017, 2.1 percent each.

[22] We also made parallel RMC calculations for female service members.

[23] Many stay/leave decisions apart from early attrition are made in these age ranges. Personnel staying beyond ten or 12 YOSs, typically at higher ages, are drawn forward by retirement benefits, resulting in high retention to 20 YOSs. In 2015, 72 percent of enlisted members were age 30 or younger and 43 percent of officers were ages 26 to 35 (Office of the Deputy Assistant Secretary of Defense for Military Community and Family Policy, 2016).

Figure 2.3
Civilian Wages for High School Graduate Men and Median Regular Military Compensation for Army Enlisted, Ages 23 to 27, Calendar Years 2000 to 2016, in 2015 Dollars

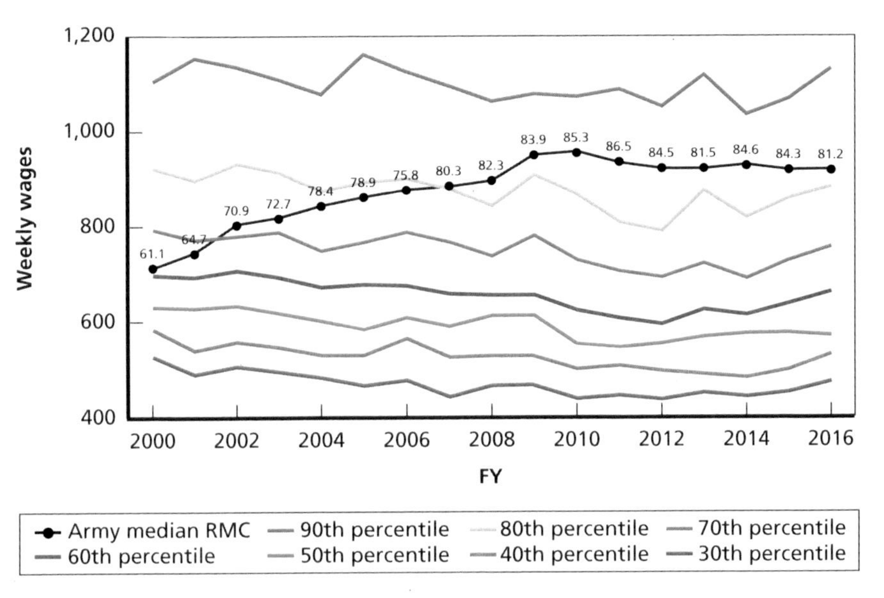

SOURCE: Active-duty pay files from DMDC; U.S. Census Bureau, 2018.
NOTE: The reference population is men ages 23 to 27 who reported high school completion as their highest level of education, worked more than 35 weeks in the year, and usually worked more than 35 hours per week. We computed the weekly wage by dividing annual earnings by annual weeks worked. The colored lines depict the wages at the indicated percentiles of the wage distribution for this population. For instance, at the 70th percentile, 30 percent of the population had higher wages and 70 percent had lower wages. The black line depicts median RMC for Army enlisted between ages 23 and 27. The numbers above the RMC line are the percentiles at which RMC stood in the population's wage distribution.

(20 percent in 2016). In 2009, 24 percent of enlisted ages 28 to 32 had five or fewer YOSs (28 percent in 2016). The presence of a significant fraction of members with low YOS numbers causes median RMC to be lower than if it were limited to service members with, say, eight to 12 YOSs and in the age range.

As seen in Figures 2.3 through 2.6, RMC increased from 2000 to 2010 and then stayed roughly constant to 2016. The increase was driven by several factors: basic-pay table restructuring that took effect in 2001

Figure 2.4
Civilian Wages for Men with Some College and Median Regular Military Compensation for Army Enlisted, Ages 28 to 32, Calendar Years 2000 to 2016, in 2015 Dollars

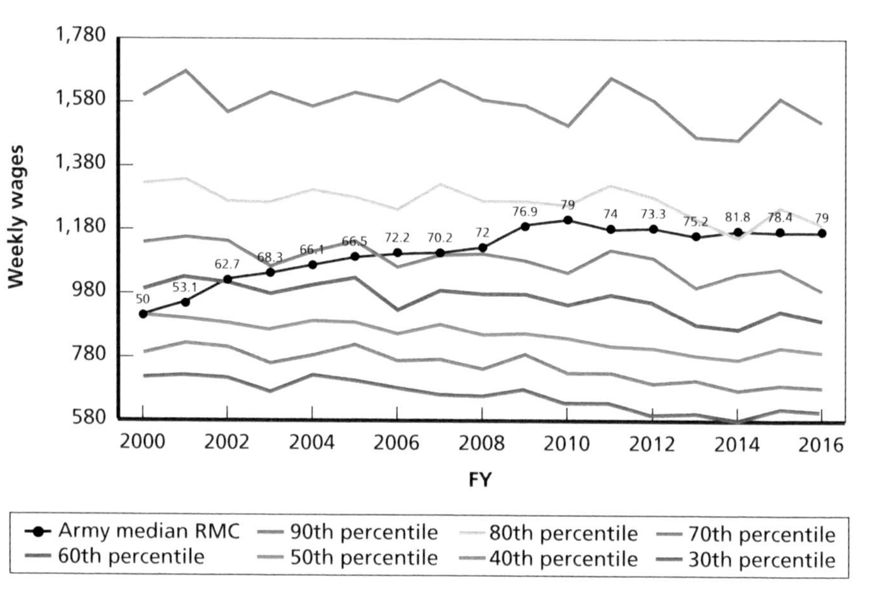

SOURCES: Active-duty pay files from DMDC; U.S. Census Bureau, 2018.
NOTE: The reference population is men ages 28 to 32 who reported some college as their highest level of education, worked more than 35 weeks in the year, and usually worked more than 35 hours per week. We computed the weekly wage by dividing annual earnings by annual weeks worked. The colored lines depict the wages at the indicated percentiles of the wage distribution for this population. For instance, at the 70th percentile, 30 percent of the population had higher wages and 70 percent had lower wages. The black line depicts median RMC for Army officers between ages 28 and 32. The numbers above the RMC line are the percentiles at which RMC stood in the population's wage distribution.

through 2003, higher-than-usual basic-pay increases from FY 2000 to FY 2010, increases implemented in the first part of the decade in BAH to cover the full cost of housing, and increases in housing cost that resulted in further BAH increases. The NDAA for FY 2000 mandated basic-pay increases equal to the percentage change in the ECI plus 0.5 percent from 2000 through 2006. Basic-pay increases continued higher-than-ECI increases through 2010 except for 2007. The BAH increases decreased the expected out-of-pocket costs for housing from

Figure 2.5
Civilian Wages for Men Who Were Four-Year College Graduates and Median Regular Military Compensation for Army Officers, Ages 28 to 32, Calendar Years 2000 to 2016, in 2015 Dollars

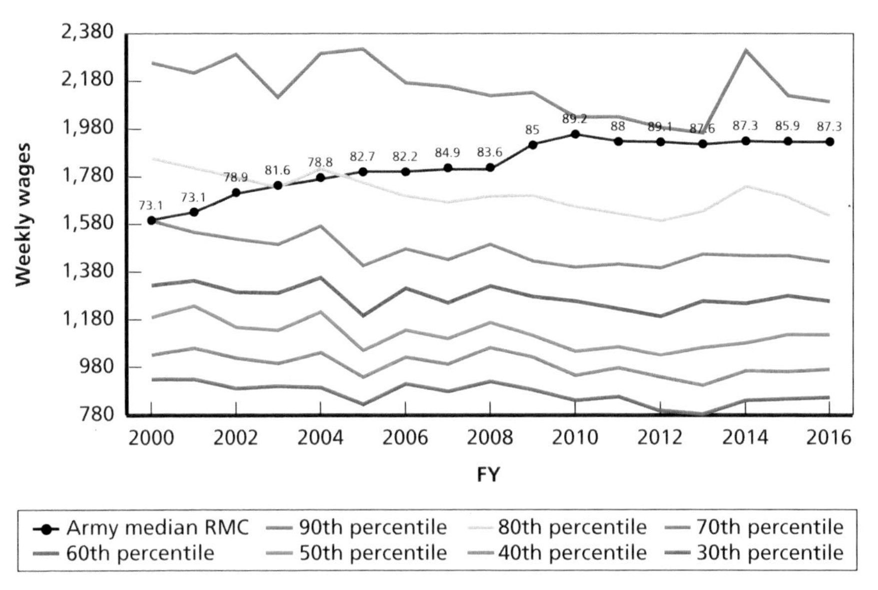

SOURCES: Active-duty pay files from DMDC; U.S. Census Bureau, 2018.
NOTE: The reference population is men ages 28 to 32 who reported bachelor's degrees as their highest levels of education, worked more than 35 weeks in the year, and usually worked more than 35 hours per week. We computed the weekly wage by dividing annual earnings by annual weeks worked. The colored lines depict the wages at the indicated percentiles of the wage distribution for this population. For instance, at the 70th percentile, 30 percent of the population had higher wages and 70 percent had lower wages. The black line depicts median RMC for Army officers between ages 28 and 32. The numbers above the RMC line are the percentiles at which RMC stood in the population's wage distribution.

20 percent in 2000 to 0 percent in 2005 (Defense Travel Management Office, 2018).[24] From 2000 to 2014, BAH grew by 63 percent, much more than the 12-percent increase in basic pay, both in 2014 dol-

---

[24] Defense Travel Management Office, 2015, explains,

> BAH rates are set based on the median housing cost for each grade and housing profile. For a given individual, an out-of-pocket expense may or may not be incurred based on the actual housing choice. If a member rents at or above the median rate for the grade/profile, that member incurs out-of-pocket expenses. The opposite is true for an indi-

Figure 2.6
Civilian Wages for Men with Master's Degrees or Higher and Median Regular Military Compensation for Army Officers, Ages 33 to 37, Calendar Years 2000–2016, in 2015 Dollars

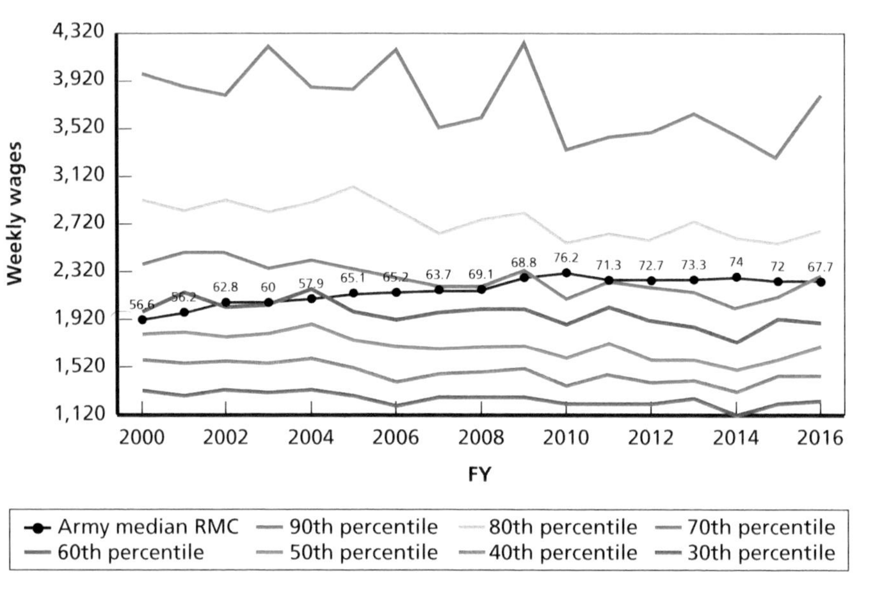

SOURCES: Active-duty pay files from DMDC; U.S. Census Bureau, 2018.
NOTE: The reference population is men ages 33 to 37 who reported master's degrees or higher as their highest levels of education, worked more than 35 weeks in the year, and usually worked more than 35 hours per week. We computed the weekly wage by dividing annual earnings by annual weeks worked. The colored lines depict the wages at the indicated percentiles of the wage distribution for this population. For instance, at the 70th percentile, 30 percent of the population had higher wages and 70 percent had lower wages. The black line depicts median RMC for Army officers between ages 33 and 37. The numbers above the RMC line are the percentiles at which RMC stood in the population's wage distribution.

lars. Looking ahead, we saw that the out-of-pocket expense for housing "will increase by one percent annually until it is capped at 5%. Thus, out-of-pocket expenses were planned to be 2% in 2016, 3% in 2017, 4% in 2018 and 5% in 2019" (Defense Travel Management Office, 2018, p. 7).

---

vidual who chooses to occupy a less expensive residence. Only a member whose housing costs are below the median will have no out-of-pocket expenses.

The increase in RMC percentile also came from a downward trend in civilian wages that leveled off around 2012 and tended to increase after 2013. In Appendix A, we estimate the trend in real civilian wages from 2000 through 2014. We found decreases of 1.14 percent per year for 23- to 27-year-old male high school graduates, 0.79 percent per year for same-age men with some college, 0.77 percent per year for 28- to 32-year old men with bachelor's degrees, and 0.45 percent per year for 28- to 32-year-old men with master's degrees or higher.

From 2010 onward, the figures indicate that, for

- enlisted members ages 23 to 27 and high school graduates, RMC has been between the 81st and 85th percentiles
- enlisted members ages 28 to 33 and with some college, RMC has been between the 75th and 82nd percentiles
- officers ages 28 to 32 who were four-year college graduates, RMC has been between the 85th and 90th percentiles
- officers ages 33 to 37 with master's degrees or higher, RMC has been between the 68th and 76th percentiles.

Summary

When comparing RMC to civilian wages, it is critically important to control for education. We computed RMC percentiles separately for high school, some college, associate's, bachelor's, and master's or higher for enlisted personnel and for bachelor's and master's or higher for officers; we also computed overall RMC percentiles. The 9th QRMC recognized the high percentage of enlisted with some college and recommended that RMC be at around the 70th percentile for that level of education. For 2016, we found an average RMC percentile of 93 for enlisted with high school and 87 for enlisted with some college, and an overall RMC percentile (for all levels of education) of 84. For officers, we found an average RMC percentile of 86 for those with bachelor's degrees and 70 for those with master's degrees or higher. The overall RMC percentile for officers is 77.

We also computed RMC percentiles for 2009. Our RMC percentiles are similar but somewhat lower than those of the 11th QRMC, and we attribute this to methodological differences.

Educational attainment has continued to increase for both enlisted and officers, and this can affect RMC percentile comparisons. When we controlled for education by using the 2016 education distribution in computing the 2009 percentiles, enlisted RMC in 2009 was at the 82nd percentile for zero to 20 YOSs and at the 81st percentile for zero to 30 YOSs. Officer RMC was at the 77th percentile for YOSs 0 to 20 and at the 76th percentile for YOSs 0 to 30—the same values as for 2016. Thus, when we held education constant, RMC increased slightly for enlisted from 2009 to 2016 and remained the same for officers.

Trend analysis shows an increase in RMC percentile from 2000 to 2016 for various age/education groups. This reflects the relatively fast military-pay growth from 2000 to 2010, as well as a downward trend in real civilian wages. The average annual decrease in civilian wages ranged from 0.5 percent per year to more than 1 percent per year.

CHAPTER THREE

Recruit Quality and Military and Civilian Pay

We considered whether the increase in RMC since 1999, when the 9th QRMC was underway, was associated with an increase in recruit quality. To address this question, we estimated reduced-form regression models on two recruiting outcomes, both of which are for accessions who have no prior service; in this chapter, we refer to them as non–prior service (NPS) accessions. The outcomes are the *recruiting rate* for HSDGs and the *non-HSDG share of accessions*. We defined the outcomes separately for *each* AFQT score category (I, II, IIIA, and IIIB). This departed from past studies that focus on high-quality accessions collectively and allowed us to detect whether the responses to military and civilian pay differed by AFQT category.[1]

The recruiting rate is the ratio of HSDG accessions in a given AFQT category to the youth high school completers in that category,

---

[1] We focus on HSDGs because research has shown attrition to be 1.5 to 2 times higher for non-HSDG than for HSDG (e.g., Buddin, 2005). However, in DoD usage, high-quality accessions are defined as I–IIIA and tier 1, where

> Tier 1 accessions are primarily high school diploma graduates, but they also include people with educational backgrounds beyond high school, as well as those with adult education diplomas, one semester of college, and in recent years, some home-schooled and virtual/distance learning graduates. (Office of the Under Secretary of Defense for Personnel and Readiness, 2016, Table D.7)

Thus, tier 1 accessions are predominantly, but not entirely, HSDGs. In this chapter, HSDGs are defined from codes 31 through 65 in MEPS data, which are for high school diploma, completed one semester of college, associate's degree, professional nursing diploma, bachelor's degree, master's degree, post–master's degree, first professional degree, doctorate degree, and postdoctorate degree. We do not include home-schooled or virtual or distance learners in HSDGs in this chapter.

net of those going on to complete four or more years of college. The share of non-HSDG accessions is the ratio of non-HSDG accessions in a given AFQT category to the total number of accessions in that category (HSDG and non-HSDG).

We first present graphics on recruit quality and describe the variables that we included in the models. We then discuss the models and results. The data period for analysis was 1999 through 2015. To indicate the magnitude of the responses to military and civilian pay, we used the estimated models to predict the recruiting outcomes for 1999 and 2015, the respective low and high years of the RMC/wage ratio in our data, controlling for other variables.

Trends in Recruit Quality

Figures 3.1 through 3.3 show how NPS recruit quality changed between 1999 and 2015. Figure 3.1 shows the percentage of accessions who are high quality, by service, defined here as HSDGs in AFQT categories I through IIIA. Figure 3.2 shows the percentages who are HSDGs, while Figure 3.3 shows the percentages who are in AFQT categories I through IIIA.

Recruit quality increased between 1999 and 2015 for the Air Force, Navy, and Marine Corps but not the Army. The Air Force, Navy, and Marine Corps increased their percentages of accessions who were high quality (Figure 3.1) and had, or reached, a very high percentage of accessions who were NPS HSDGs (Figure 3.2). The Army's percentage of accessions who were high quality fell after 2004, rebounded to its initial level, then stayed there. Its HSDG percentage bottomed out in 2007 and then rose to a high level near that of the other services. Its percentage accessions in categories I through IIIA in the active component declined fairly steadily after 2004 (Figure 3.3), while this percentage increased in the other services.

Figure 3.1
Percentage of Active-Component Non–Prior Service Accessions Who Were Category I–IIIA High School Diploma Graduates, by Service

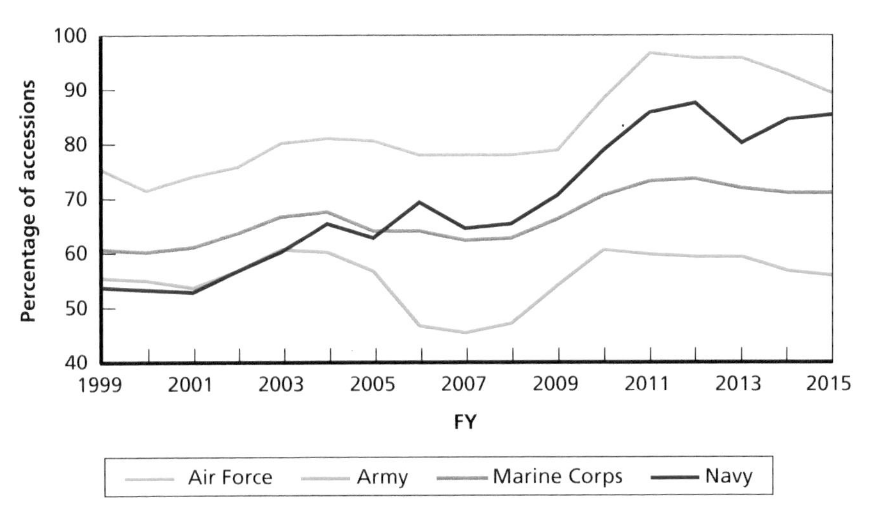

SOURCE: Office of People Analytics, undated.
NOTE: An HSDG is someone with a high school diploma, associate's degree, professional nursing diploma, bachelor's degree, master's degree, post–master's degree, first professional degree, doctoral degree, postdoctorate degree, or one semester of college completed. Category I–IIIA personnel are those who scored in the upper half of the AFQT score distribution.

Explanatory Variables

The explanatory variables used in the regression model were military and civilian pay, recruiting goal, deployment, unemployment, gender, and a post-2009 indicator (that is, an indicator of being after 2009; see "Post-2009 Indicator" later in this section). We introduce them here to describe their variation and provide context for the regression analysis.

Military and Civilian Pay

To construct a variable for military and civilian pay, we used RMC for an E-4 with four YOSs and the median civilian wages of 18- to 22-year old male and female workers with high school education (and not more). We considered two pay measures: the RMC percen-

Figure 3.2
Percentage of Active-Component Non–Prior Service Accessions Who Were High School Diploma Graduates, by Service

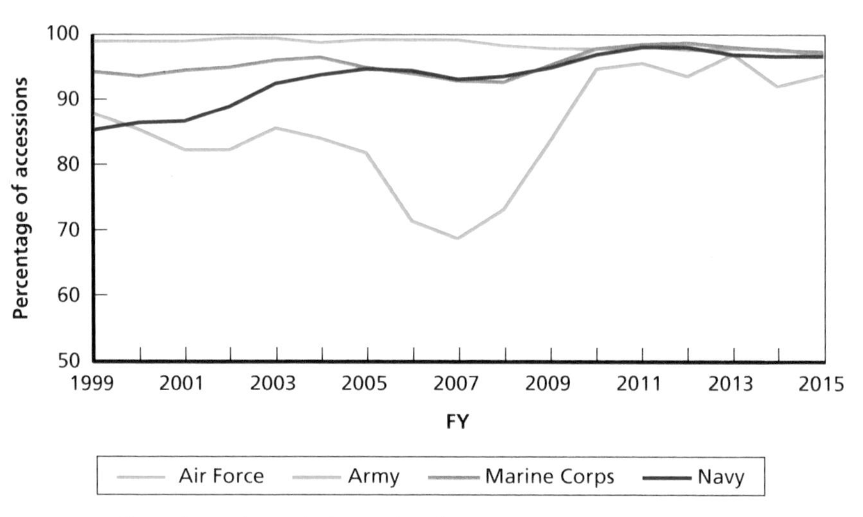

SOURCE: Office of People Analytics, undated.

tile and the military/civilian wage ratio, and we chose the latter. The latter approach is consistent with the approach used in past studies (e.g., Simon and Warner, 2007; Asch, Heaton, et al., 2010), and using the former approach has the disadvantage that pay must increase by a larger absolute amount to move from, say, the 85th to the 90th percentile than from, say, the 70th to the 75th, while the pay ratio has the same interpretation throughout its range. That said, the regression results using the pay ratio were nearly the same as the RMC percentile results and had similar statistical significance. RMC is from the Greenbook, and the wage is from the March CPS.

Figure 3.4 displays the RMC percentile, smoothed to adjust for variation resulting from sample size.[2] As expected, the RMC percentile increased from 1999 to 2015 for men and women. Its rate of increase was higher in the first years then decreased. This is because military-pay growth slowed after 2010 and because of the log-normal nature of

[2] Appendix A describes the smoothing method and contains the raw and smoothed values.

Figure 3.3
Percentage of Active-Component Non–Prior Service Accessions Who Were in Categories I Through IIIA, by Service

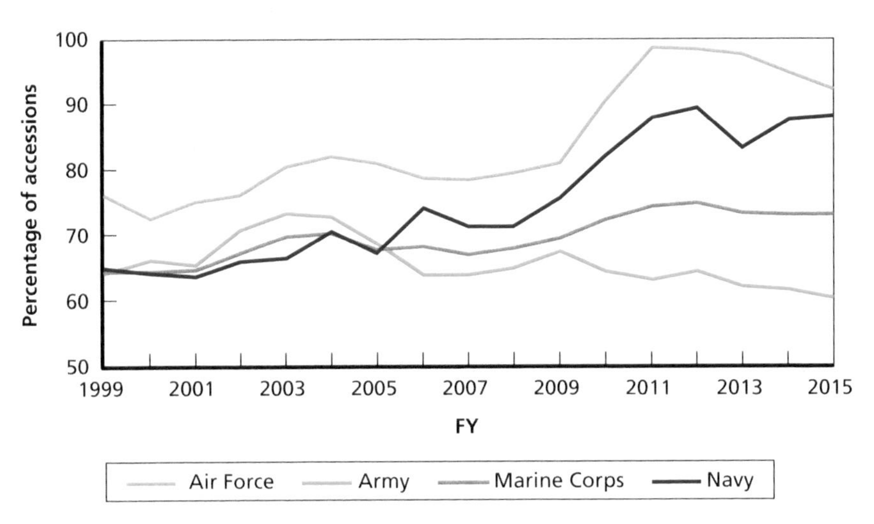

SOURCE: Office of People Analytics, undated.

the wage distribution. In the upper tail of the distribution, RMC must increase by an ever-greater absolute amount to move up 1 percentile. The RMC/wage ratio does not have this issue. As seen in Figure 3.5, the RMC/wage ratio increased steadily throughout the period. The figure shows the raw ratio and a linear curve fitted to it.[3]

Recruiting Goal

The Army had the largest recruiting goal, roughly double that of each of the other services in the middle of the decade (Figure 3.6). The recruiting goal that is set depends on a service's authorized end strength and outflow of personnel. James Dertouzos developed a theoretical enlistment supply model in which recruiter effort increases as recruiting goals increase, holding other factors affecting recruiting constant, such as recruiting resources (including bonuses and adver-

---

[3] A quadratic curve can also be fitted. The quadratic coefficient is not significant for women but is significant for men. In our regressions, we used the raw ratio.

Figure 3.4
Smoothed Regular Military Compensation Percentiles: Male and Female High School Graduates, Ages 18 to 22

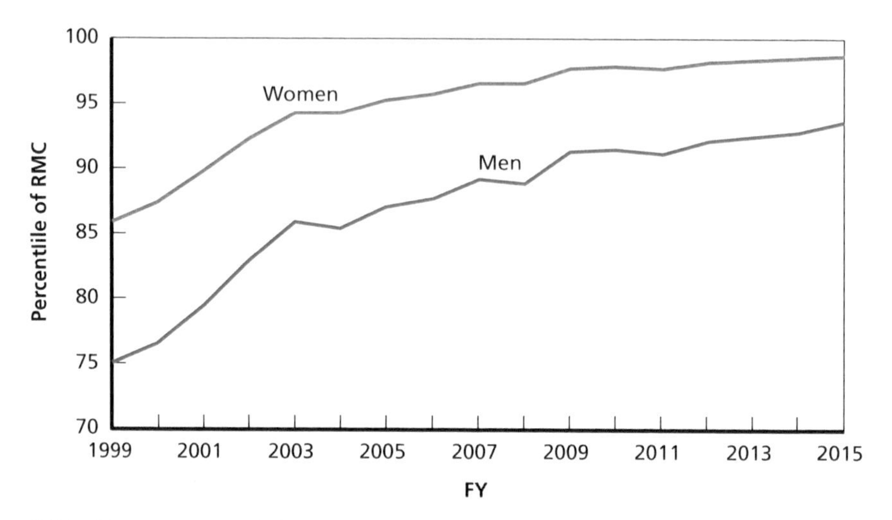

SOURCES: We estimated the curves as described in Appendix A. The data used in the regressions were based on our tabulations of RMC as a percentile of the civilian wages of male and female high school graduates ages 18 to 22. These tabulations used CPS data and DMDC pay files.
NOTE: RMC is for an E-4 with four YOSs. The RMC percentile is relative to the wage distribution for 18- to 22-year-old workers with high school (and not additional education) who had more than 35 hours of work in the year and more than 35 usual weekly hours of work.

tising) (Dertouzos, 1985). Past research, including the Dertouzos study, supports this hypothesis (Dertouzos, 1985; Simon and Warner, 2007; Asch, Heaton, et al., 2010). The recruiting goals decreased on net during this period, but the timing of the decrease differed by service. The Army goal decreased from 2008 onward, while the Navy goal decreased from 2000 to 2006. The Air Force goal trended down steadily, apart from a temporary drop in 2005. The Marine Corps goal, like the Army's, decreased after 2008.

Deployment

The recruiting environment changed dramatically after 9/11. Operations in Iraq and Afghanistan greatly increased Army and Marine

Figure 3.5
RMC/Median Wage Ratio: Male and Female High School Graduates, Ages 18 to 22

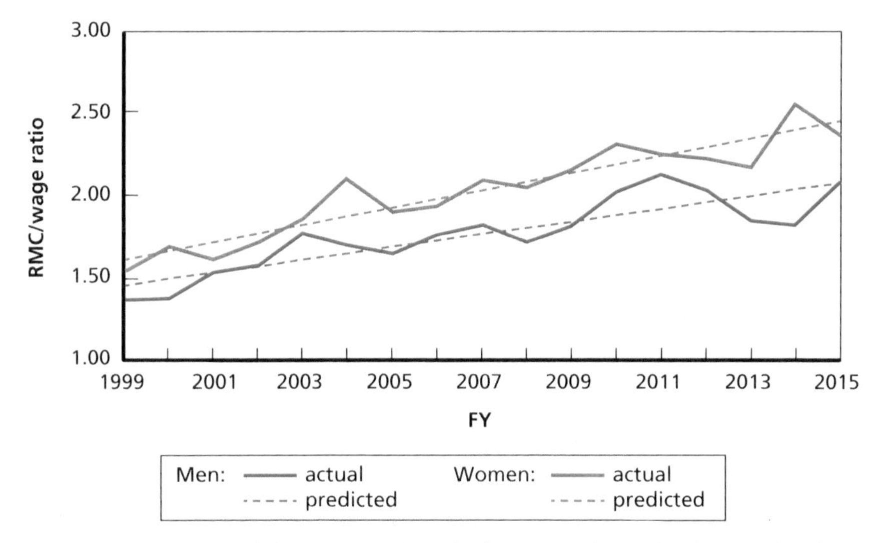

SOURCES: We estimated the curves as described in Appendix A. The data used in the regressions were based on our tabulations of RMC as a percentile of the civilian wages of male and female high school graduates ages 18 to 22. These tabulations used CPS data and DMDC pay files.
NOTE: RMC is for an E-4 with four YOSs. The RMC percentile is relative to the wage distribution for 18- to 22-year-old workers with high school (and not additional education) who had more than 35 hours of work in the year and more than 35 usual weekly hours of work.

Corps deployments, and Middle East theaters were designated imminent-danger or hostile-fire areas. We used the number of personnel receiving imminent-danger or hostile-fire pay to measure the extent of deployment. With 1999 = 1.00, the Army and Marine Corps had six to nine times more deployed personnel in 2003 through 2010 than in 1999 (Figure 3.7). Navy and Air Force increases were 0.3 to 2.5 times higher than in 1999.

Researchers in past studies have found that operations in Iraq and Afghanistan had a negative effect on recruiting, although the studies varied in terms of the estimated size of the effect (Simon and Warner, 2007; Asch, Heaton, et al., 2010; Christensen, 2017). Exten-

Figure 3.6
Recruiting Goals, by Service

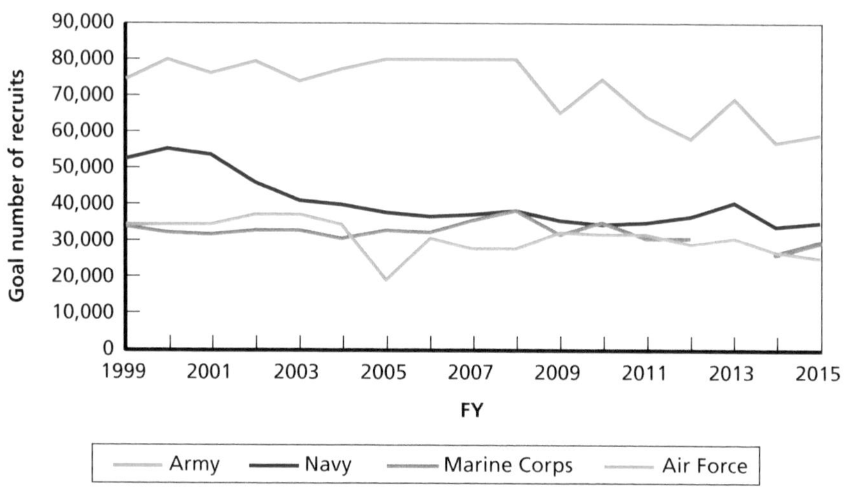

SOURCE: Accession Policy Directorate, 2017.

sive deployments between 2002 and 2009 might have made it more difficult for the Army to enlist high-quality recruits. The percentage of Army accessions who were HSDGs dropped in the middle of the decade (Figure 3.2), which also meant a fall in high-quality accessions (Figure 3.1). Marine Corps deployment also rose, but the service's HSDG and high-quality recruiting rose.

Unemployment

The trend in recruit quality shown in Figures 3.1 through 3.3 shows that the percentage of recruits who were high-quality recruits and who were in categories I through IIIA rose rapidly from 2009 to 2012, a period when the unemployment rate was high (Figure 3.8). The unemployment rate was low in the late 1990s, during the dot.com boom that ended in the recession of 2000–2001. After the economy recovered, the unemployment rate declined until the Great Recession in 2008, and the unemployment rate doubled that year and did not return to its prerecession level until 2015. The unemployment rates for young men and women, ages 16 to 24, are higher than the overall rate

Figure 3.7
Enlisted Personnel Receiving Imminent-Danger or Hostile-Fire Pay

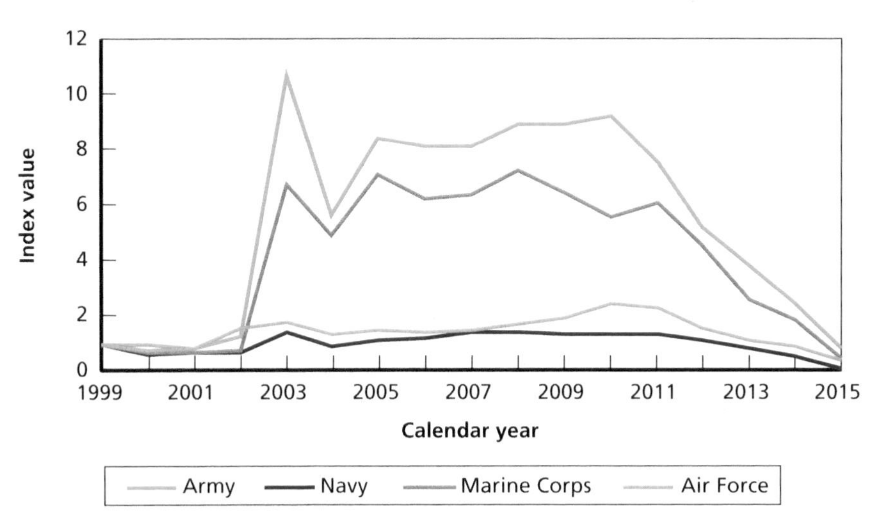

SOURCES: Military-pay files.
NOTE: 1999 = 1.00.

but vary in parallel with it. Researchers in past studies have found that the number of high-quality enlistments in each service, including the Army, is positively associated with the civilian unemployment rate (for reviews of past studies, see, e.g., Asch, Heaton, et al., 2010, and Asch, Hosek, and Warner, 2007).

Post-2009 Indicator

The Post-9/11 Veterans Educational Assistance Act of 2008 (also known as the Post-9/11 GI Bill) for education benefits took effect in August 2009 (Pub. L. 110-252, Title V, 2008). The bill expanded benefits by offering to cover tuition at a level equal to the tuition of a service member's home-state four-year public university, plus offering BAH while attending school. Prior to that point, the services had the same education benefit and the Army offered a supplement to it, the Army College Fund, for high-quality recruits in critical occupations. The Post-9/11 GI Bill made benefits available to recruits of all services on equal terms, regardless of quality, and, given the generosity of the

Figure 3.8
Civilian Unemployment Rate, 1999 to 2015

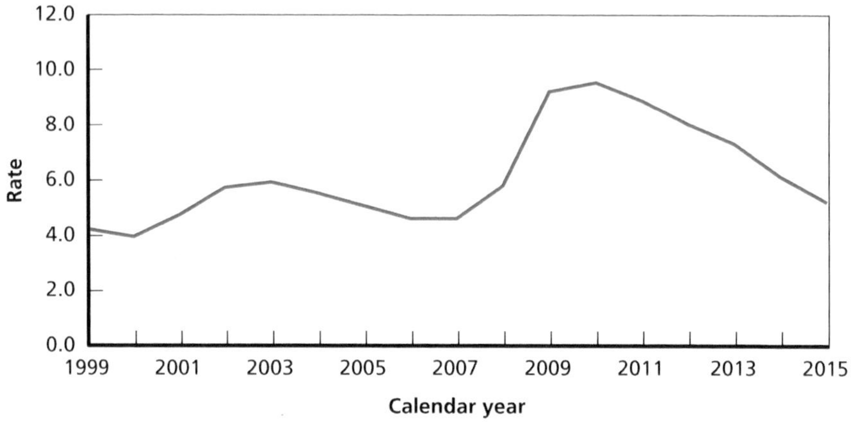

SOURCE: Bureau of Labor Statistics, undated.

benefits, the Army lost the recruiting advantage it had with the Army College Fund. Thus, because of the Post-9/11 GI Bill, we included a post-2009 indicator variable for the years 2010 through 2015.

Modeling the Relationship Between Recruiting Rate and RMC/Wage Ratio

Recruiting is determined by supply and demand behavior—a person's willingness to join a service and a service's willingness to enlist the person, given willingness. This is true throughout the recruitment process (e.g., prospecting, making contact, signing an enlistment contract, time in the delayed-entry program [DEP],[4] and, finally, accessing). The model we estimated is a reduced-form model that reflects supply and demand influences but does not identify the effects of the supply and demand variables separately.

[4] In the Army DEP is called the Future Soldiers Program. We use the term *DEP* generically to apply to all of the services' programs.

We view willingness to enlist as a variant of the random utility model (McFadden, 1983).[5] An individual's willingness to join a service depends on military pay relative to civilian pay, job opportunities as measured by the unemployment rate, job and school opportunities related to AFQT, the chance of being deployed in hostile operations, and possible differences in preferences and opportunities related to gender. It also depends on factors not observed in our data, including the military occupational specialties that are offered, bonuses, educational benefits, ship date, information from advertising or service websites, the influence of family and friends, and aspects of military service, such as its roles, missions, tradition, and values (see, e.g., Eighmey, 2006).

On the demand side, the recruiting command wants to meet a quantity goal and meet or exceed a quality goal. The service goals must comply with DoD guidance that calls for at least 60 percent of accessions to be in categories I through IIIA and at least 90 percent to be tier 1 (Kapp, 2013; Sellman, 2004). A tier 1 recruit is a recruit with a traditional high school diploma, a semester of college, or an associate's, bachelor's, or graduate degree, and tier 1 recruits are predominantly HSDGs. Category I–IIIA HSDG recruits count toward the quantity and quality goals, while category IIIB HSDG and category IV HSDG recruits count only toward the quantity goal. This suggests that the probability of an HSDG's recruitment conditional on the person being willing to enlist relates to a preference ordering:

$$\Pr\left(accept \mid willingCatsIthroughIIIA\right) = 1$$
$$\Pr\left(accept \mid willingCatIIIB\right) \leq 1$$
$$\Pr\left(accept \mid willingCatIV\right) \ll 1.$$

The "equal" part of "less than or equal" for IIIB allows for the possibility that meeting the quantity goal might require accepting all willing IIIBs. It also reflects the possibility of identifying high-potential

[5] Many reports (e.g., Kilburn and Klerman, 1999) discuss the enlistment-decision model.

IIIB prospects through nontraditional methods.[6] Thus, the AFQT can affect a service's willingness to accept someone who is willing to join, especially if the person is at IIIB or less. Also, a person's willingness can depend on the AFQT, as mentioned, because it can be related to job prospects, college expectations, and college opportunities. For these demand and supply reasons, our analysis allowed for possible differences by AFQT category in the recruiting rate's responsiveness to the RMC/wage ratio.

Our model is a reduced-form model because it does not identify structural equations for the demand and supply sides. For instance, unemployment can increase willingness to enlist, and, in response, the services can decrease their recruiting resources, such as recruiters, advertising, and bonuses, and tighten eligibility.[7] Our reduced-form unemployment coefficient is the net effect of these responses. The direct effect of unemployment on supply is positive, but, if unemployment triggers a large enough decrease in recruiting resources at the same time, the coefficient on unemployment in the reduced-form

[6] The Army has explored nontraditional measures of quality in its Assessment of Recruit Motivation and Strength (ARMS), Tier 2 Attrition Screen, and Tailored Adaptive Personality Assessment System (TAPAS) programs. The Army implemented the ARMS program in 2005 and suspended it in 2009. It permitted waivers to applicants who exceeded weight and body fat standards provided they passed the ARMS test (Loughran and Orvis, 2011; Bedno et al., 2010). Recruits entering under the ARMS program proved to be no more likely to attrit than non-ARMS recruits were. Between 2006 and 2008, the Tier 2 Attrition Screen (TTAS) allowed for more tier 2 contracts (Heffner, Campbell, and Drasgow, 2011). TTAS was the forerunner to TAPAS, which aims to identify prospects with desirable attributes, such as commitment, drive, personality, and leadership, that are not measured solely by the Armed Services Vocational Aptitude Battery (ASVAB) or the AFQT, which is based on a subset of ASVAB scores (Nye et al., 2012). Initial operational test and evaluation of TAPAS began in 2009 on several large occupational specialties, including infantry, military police, combat medics, and motor transport operators. TAPAS is not currently being used operationally in recruiting, but its development and evaluation continues (Drasgow, 2013; Stark et al., 2014).

[7] For instance, the Army grants various waivers and can tighten them. The largest categories involve conduct and medical waivers. Conduct waivers might be for traffic or, primarily, nontraffic misdemeanors, although, in the past, a limited number of felony waivers were also granted. Medical waivers can be for height, weight, or, the lion's share, for a medical condition requiring a waiver to enlist (Orvis, Maerzluft, et al., 2018). The normal rate of medical waivers in the Army is on the order of 8 to 10 percent of accessions.

model could be negative. Similarly, deployment might have a positive or negative effect on supply, and, if the supply effect is negative, the service might, for instance, increase its enlistment bonuses (as mentioned above), nullifying the negative effect.[8]

It would be ideal to identify structural (causal) effects of recruiting resources, which are endogenous to the recruiting process, but doing so requires exogenous variation in explanatory variables, such as through enlistment bonus, recruiter, or advertising experiments, or through the use of instrumental variables. Our data are not from experiments, and we did not have instruments or enough data to estimate instrumental-variable models.[9] Instead, the variables we included in the reduced-form model were outside the control of the recruiting command and, as suggested, are external drivers to its resourcing decisions. These variables are military and civilian pay, recruiting goal, deployment, unemployment, and eligibility. Although the reduced-form model does not identify causal effects, such as the causal effect of pay, it avoids issues of bias that would have arisen if we had included observed bonuses, recruiters, and advertising.

Still, poor recruiting and retention conditions in one year might result in a higher-than-expected military-pay raise and a higher recruiting goal in the next. These conditions—autocorrelated errors and policy actions affecting pay and retention—could bias downward the RMC/wage and recruiting goal coefficients and produce low standard

---

[8] As deployments rose, the Army and Marine Corps paid more enlistment bonuses. The correlation between the number of enlisted personnel receiving imminent-danger pay and the number of recruits receiving enlistment bonuses between 1999 and 2015 was 0.41 for the Army, 0.05 for the Navy, 0.49 for the Marine Corps, and –0.29 for the Air Force. The negative correlation for the Air Force is counterintuitive. It results from a large decrease in 2005 in the number of accessions receiving enlistment bonuses. The correlation from 2005 through 2015 is 0.13, a positive value, as expected. Enlistment bonuses helped the Army meet its recruiting objectives during the high-deployment years. Beth Asch and her colleagues estimated that, without the use of bonuses, the Army would have enlisted 20 percent fewer high-quality recruits between 2004 and 2008 (Asch, Heaton, et al., 2010). Despite the similarity between Army and Marine Corps correlations, the Marine Corps made little use of enlistment bonuses until 2007.

[9] An instrument for bonuses, for instance, is correlated with bonuses—hence useful to predict bonuses—but uncorrelated with the error term in the recruiting outcome equation.

errors (and thus high *t*-statistics). A downward bias would imply that the coefficients are conservative estimates of the true effect.[10]

We ran separate models for each service. The dependent variable is the logit of the recruiting rate. There is a recruiting-rate observation for each AFQT category for men and another for women. The RMC/wage coefficient is allowed to differ by category. Also, the intercepts are allowed to differ by AFQT category interacted with gender. The coefficients for recruiting goal, deployment, unemployment, and the post-2009 indicator are the same across AFQT categories. In the logit specification, the *percentage* change in the recruiting rate with respect to a continuous variable equals the coefficient times one minus the recruiting rate (i.e., $\beta[1-p]$).[11] The recruiting rates are low percent-

[10] We caution that some of the estimated relationships might be spurious because one or more terms in the regression is nonstationary. Udny Yule demonstrated that spurious correlations might be detected between two nonstationary time series, although the underlying series themselves have no relationship (Yule, 1926). Granger and Newbold also demonstrated this using a Monte Carlo analysis (Granger and Newbold, 1974). Unfortunately, the time series we used were too short to be able to completely eliminate the possibility that one or more of the series is nonstationary.

[11] In the logit regression specification, the marginal change in p with respect to a continuous explanatory variable x is

$$\frac{\partial p}{\partial x} = \beta(1-p)p.$$

Therefore, the percentage change in p with respect to x is

$$\left(\frac{\partial p}{\partial x}\right)\left(\frac{1}{p}\right) = \beta(1-p).$$

We also ran regressions in which the dependent variable was the log of the recruiting rate. In that specification, the coefficients represent the percentage change in the dependent variable for a one-unit change in the explanatory variable. The results were virtually the same as $\beta(1-p)$ from the logit.

ages, so $1 - p \approx 1$ and the percentage change is roughly equal to the coefficient itself, β.[12]

A final point is to observe that one might expect the results to differ for the Army and the other services because the Army's recruiting goal is the largest, being roughly twice that of each of the other services. However, the extent to which the magnitude of the recruiting goal makes a difference depends on resourcing decisions that the services made and the quality they required. The marginal cost of a recruit of a given quality might be higher for the Army if it must go deeper into the population of prospective recruits; yet, by programming enough resources to recruiting, the Army might attain the same quality as the other services. But resources have other uses, and the same quality might not be needed. The issue, then, turns on the expected benefit from higher-quality recruits relative to their cost and, further, whether the positions in the Army need to be manned by the same quality of recruit, on average, as needed in the other services. In short, service differences in recruiting cost and required quality could give rise to differences in recruit quality and—relevant to this research—differential responses in recruit quality by service when the rise in RMC is the same across services.

Modeling the Relationship Between the Share of Non–High School Diploma Graduate Accessions and Regular Military Compensation/Wage Ratio

Here, we focus on the share of accessions who are non-HSDGs, again by AFQT category. The intuition is that the service will prefer an

[12] For indicator variables, the impact on the recruiting rate required evaluating

$$\frac{e^{\beta X}}{1+e^{\beta X}}$$

for the variable at 1 versus 0, with other explanatory variables held at some level (e.g., their means).

HSDG recruit to a non-HSDG recruit, other things being equal.[13] Yet, if recruiting conditions are more difficult than expected, recruiting outcomes are below goal, goals have unexpectedly increased, or recruiting stations become short-staffed, recruiting a non-HSDG might be more attractive. This would be the case if, given the adverse recruiting conditions, the marginal benefit from meeting the quantity goal relative to the marginal cost of doing so were greater for a non-HSDG than an HSDG recruit. This intuition extends beyond unexpected, difficult recruiting conditions, though. Judging by recruiting outcomes (Figures 3.1 through 3.3), non-HSDG accessions are a regular part of the recruiting mix that the services program resources to obtain, although the actual outcomes will depend on conditions realized during the recruiting year. In 1999, more than 10 percent of Army and Navy accessions were non-HSDG, and, although the percentage in recent years has been well below 10 percent for all services, in 2006, more than 30 percent of Army accessions were non-HSDGs (Figure 3.2). Overall, the cost of an all–high-quality or all-HSDG accession cohort might be too high relative to the expected value to the service. (We formalize these ideas in Appendix D.)

To capture the idea that non-HSDG accessions serve as an outlet, we estimated models of the share of accessions who were non-HSDGs. The models include the same explanatory variables as the recruiting-rate models, and the dependent variable is the logit of the share of accessions who were non-HSDGs.

Regression Results

The regression estimates for the recruiting-rate and non-HSDG share models are in Tables B.6 and B.7. Figures 3.9 and 3.10 depict the RMC/wage coefficient by AFQT category for each service for the recruiting-rate and non-HSDG share models, respectively. The RMC/

---

[13] An exception would be high-potential non-HSDG prospects identified by nontraditional testing.

Figure 3.9
Regular Military Compensation/Wage Coefficients from Logit Regressions of Recruiting Rate for Armed Forces Qualification Test Categories I Through IIIB, by Service

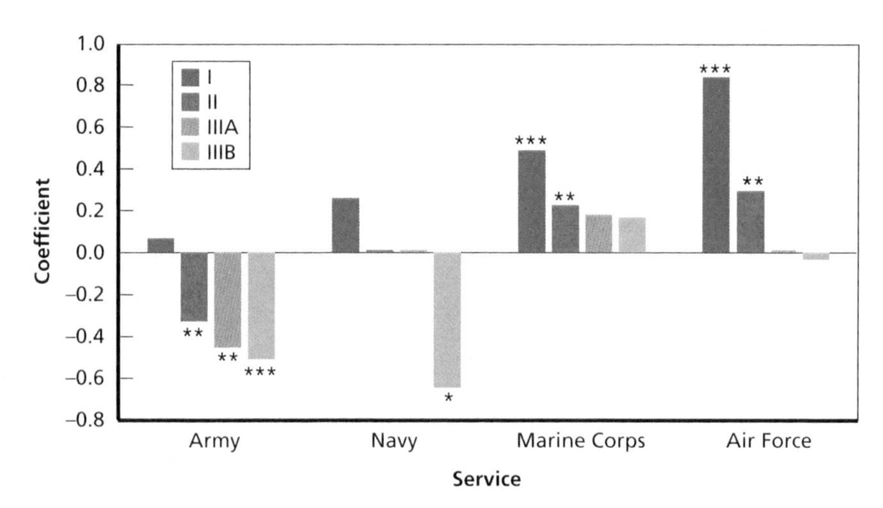

NOTE: The bars in the figure show the overall RMC percentile coefficient for each AFQT category. For example, the overall RMC percentile coefficient for category I is RMC's effect on the logit recruiting rate for AFQT category I. It equals the sum of the RMC coefficient for IIIA, the reference group, and the RMC × category I coefficient in Table B.12 in Appendix B. *** = statistically significant at the 1-percent level. ** = statistically significant at the 5-percent level. * = statistically significant at the 10-percent level.

wage coefficient indicates the relationship between the RMC percentile and the recruiting outcome variable.

Recruiting Rate

An increase in the RMC/wage ratio was associated with the following:

- *Army*: no change in the recruiting rate for category I and decreases in the rates for II, IIIA, and IIIB
- *Navy:* a decrease in the IIIB recruiting rate
- *Marine Corps:* an increase in the recruiting rates for categories I and II
- *Air Force:* an increase in recruiting rates for categories I and II.

Figure 3.10
Regular Military Compensation/Wage Coefficients from Logit Regressions of the Non–High School Diploma Graduate Share of Accessions for Armed Forces Qualification Test Categories I Through IIIB, by Service

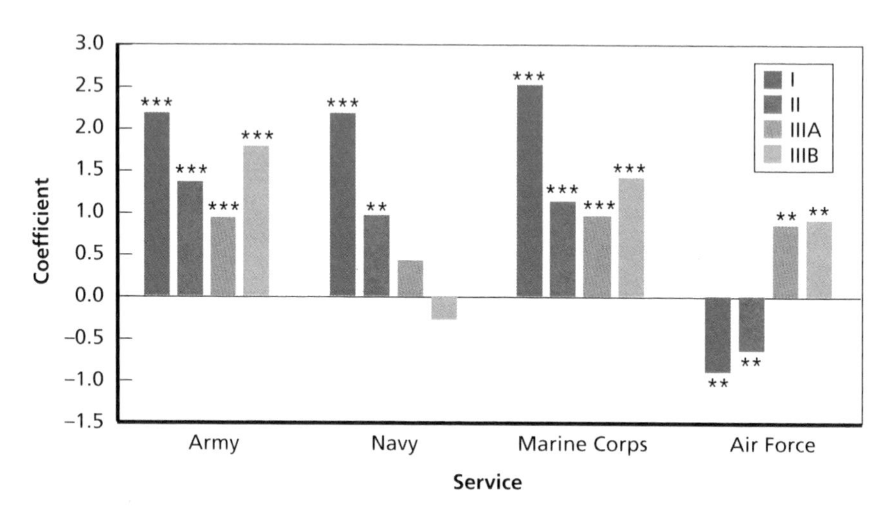

NOTE: The bars in the figure show the overall RMC percentile coefficient for each AFQT category. For example, the overall RMC percentile coefficient for category I is RMC's effect on the share of non-HSDG accessions for AFQT category I. It equals the sum of the RMC coefficient for IIIA, the reference group, and the RMC × category I coefficient in Table B.13 in Appendix B. *** = statistically significant at the 1-percent level. ** = statistically significant at the 5-percent level. * = statistically significant at the 10-percent level.

Non–High School Diploma Graduate Share

An increase in the RMC/wage ratio was associated with the following:

- *Army:* an increase in the share of non-HSDGs in all categories (I through IIIB)
- *Navy:* an increase in the share of non-HSDGs in categories I and II
- *Marine Corps:* an increase in the share of non-HSDGs in all categories (I through IIIB)
- *Air Force:* a decrease in the share of non-HSDGs in categories I and II and an increase in categories IIIA and IIIB.

Thus, with respect to the RMC/wage ratio, the Navy increased its recruit quality mix by cutting back on IIIB, and the Marine Corps and Air Force increased their quality mixes through higher category I and II recruiting rates. For the Army, RMC/wage was associated with no change in the category I recruiting rate and, contrary to what one might expect, lower recruiting rates in II and IIIA, as well as IIIB.

Further, an increase in the RMC/wage ratio was associated with a higher share of non-HSDGs in all categories in the Army and Marine Corps and in categories I and II in the Navy. Thus, in the Army, Navy, and Marine Corps, the RMC/wage ratio was associated with an increase in the *quality* of non-HSDGs in that the non-HSDG share increased in categories I, II, and IIIA. Still, the non-HSDG share in IIIB also increased in the Army and Marine Corps. The findings for the Air Force were mixed, with a decrease in this share for categories I and II but an increase in the share in IIIA and IIIB.

It might seem puzzling that the Army did not increase its recruiting rate for categories I through IIIA as the RMC/wage ratio rose, but the Army did increase its non-HSDG share of accessions in the high-AFQT categories (I, II, and IIIA). In addition, the Army increased its HSDG share after 2009 (Figure 3.2) and cut back on its non-HSDG share. The results indicate that, as it cut back on non-HSDGs, it became more selective, bringing in more high-scoring non-HSDGs as the RMC/wage ratio rose. This behavior can be explained within the context of the model in Appendix D. If non-HSDGs have a lower marginal cost of recruitment, the Army could have, by this approach, held down its recruiting cost as the RMC/wage ratio rose yet still met the DoD benchmarks of 60 percent categories I through IIIA and 90 percent tier 1. (Note that the floors are for those dimensions separately, not for category I–IIIA HSDGs jointly.)

Predicted Change in Recruiting Rate and Non–High School Diploma Graduate Share

We made two sets of predictions. We first used the estimated models to predict the recruiting rate and non-HSDG share in 1999 and 2015, using the values of the explanatory variables for those years. The predictions are for men, although the pattern of predictions is the same for women.[14] The purpose of these predictions was to show beginning- and end-of–data period predicted values, given the values of the explanatory variables at those points. The predictions would be fairly accurate because the R^2 of the recruiting-rate regressions ranged from 0.94 to 0.99, and the R^2 of the non-HSDG share regressions ranged from 0.73 to 0.86—not as good but still fairly high.

The first predictions provide context for the second set of predictions. We used the second set of predictions to isolate the RMC/wage ratio's effect on the recruiting rate and non-HSDG share. That is, the first set of predictions shows the overall change when all of the explanatory variables (for men) change; they indicate largely what one can see from the raw changes, and this is because the fit of the models (R^2) is high. The second set of predictions is important for understanding the extent to which the sizable increase in relative pay from 1999 to 2015 was associated with changes in the AFQT and HSDG/non-HSDG mixes of accessions. For 1999 and 2015, we used the measured ratios for 1999 and 2015 and fixed all other model variables at their means (or 0, in the case of indicators) to make predictions. We then calculated the percentage change in the predictions. Because this analysis varied only the RMC/wage ratio, the differences in the predictions are solely due to the changed ratio and represent the predicted impact on recruiting rate and non-HSDG share between 1999 and 2015 if nothing else changed. The values of the other variables are the means of recruiting goal, deployment, and unemployment, with the female and post-2009 indicators set to 0.[15]

---

[14] This is because the female indicators, in effect, shift the regression intercept but do not affect the contributions of the RMC/wage ratio, recruiting goal, deployment, unemployment, or post-2009 to the prediction.

[15] The nonlinear nature of the relationship between the dependent variable and the explanatory variables means that the RMC/wage ratio's predicted impact on the dependent variable depends on the point of evaluation. We could have chosen 1999 values of the other explanatory variables as the point of evaluation, for instance, or the 2015 values, and so forth, and choosing the mean values of recruiting goal, deployment, and unemployment was, we think, a reasonable compromise. Also, because we focused on predictions about men for the first

The predicted recruiting rate and non-HSDG share for 1999 and 2015 are shown in Table 3.1. The upper panel has the predicted values for 1999, the middle panel has the predicted values for 2015, and the lower panel shows the percentage change in the predicted values. The percentage changes that are statistically significant are shaded in gray. As seen, the Army recruiting rates for II and IIIA—categories in the high-quality range—were lower in 2015 than in 1999. The Navy recruiting rate for category I was higher and for IIIB was lower in 2015 than in 1999. The Marine Corps recruiting rates for categories I, II, and IIIA were higher in 2015 than in 1999, and the Air Force recruiting rate was higher for category I and lower for IIIA and IIIB. The Army and Navy shares of accessions who were non-HSDGs was lower for II, IIIA, and IIIB in 2015 than in 1999. The Marine Corps non-HSDG share was higher for categories I and II and lower for IIIB, while the Air Force non-HSDG share was higher across the board and especially for IIIA and IIIB.

Table 3.2 shows the predicted recruiting rate and non-HSDG share with the 1999 and 2015 RMC/wage ratios, and all other variables set to their mean value (or 0, in the case of indicators). We used the RMC/wage ratio for men. The RMC/wage ratio was 1.36 in 1999 and 2.08 in 2015 for men (and 1.54 in 1999 and 2.36 in 2015 for women). The lowest and highest values of the RMC/wage ratio were 1999 and 2015, respectively. Shading indicates predictions that are statistically significant (that is, significantly different from 0 at 10 percent or better; coefficient significance is shown in Tables B.6 and B.7 in Appendix B).

Focusing first on the statistically significant predictions for the recruiting rate, at these two points of evaluation for the RMC/wage ratio, we see a negative change of 20 percent and 26 percent in the Army recruiting rate for II and IIIA, respectively. We shaded the percentage changes based on RMC/wage coefficients that are statistically significantly different from 0. The IIIB recruiting rate is 29 percent lower, which is consistent with taking in fewer IIIBs at higher RMC/

---

set, we also made the predictions using the RMC/wage ratio for men (we turned off the female-indicator variables).

Table 3.1
Predicted Recruiting Rate and Non–High School Diploma Graduate Share for Men, by Armed Forces Qualification Test Category, and Percentage Change, 1999 and 2015

AFQT Category	Recruiting Rate				Non-HSDG Share			
	Army	Navy	Marine Corps	Air Force	Army	Navy	Marine Corps	Air Force
1999								
I	0.033	0.022	0.012	0.017	0.036	0.037	0.012	0.012
II	0.056	0.039	0.029	0.037	0.136	0.114	0.029	0.014
IIIA	0.078	0.048	0.042	0.046	0.243	0.196	0.042	0.005
IIIB	0.078	0.057	0.041	0.028	0.125	0.060	0.041	0.005
2015								
I	0.035	0.028	0.017	0.025	0.033	0.036	0.040	0.019
II	0.045	0.041	0.036	0.036	0.074	0.049	0.035	0.026
IIIA	0.057	0.051	0.049	0.037	0.109	0.064	0.044	0.025
IIIB	0.077	0.018	0.041	0.003	0.048	0.013	0.021	0.024
Percentage change								
I	5	26	46	44	–8	–2	237	56
II	–20	6	20	–2	–45	–57	19	84

Table 3.1—Continued

AFQT Category	Recruiting Rate				Non-HSDG Share			
	Army	Navy	Marine Corps	Air Force	Army	Navy	Marine Corps	Air Force
IIIA	–27	6	17	–20	–55	–68	5	444
IIIB	–1	–68	0	–88	–62	–79	–48	436

NOTE: Predicted from regression coefficients in Tables B.6 and B.7 in Appendix B using explanatory variables at their values for 1999 and 2015, respectively. None of these values was statistically significant.

Table 3.2
Predicted Recruiting Rate and Non–High School Diploma Graduate Share for Men Allowing Only the Regular Military Compensation/Wage Ratio to Change, by Armed Forces Qualification Test Category, and Percentage Change, 1999 and 2015

AFQT Category	Recruiting Rate				Non-HSDG Share			
	Army	Navy	Marine Corps	Air Force	Army	Navy	Marine Corps	Air Force
1999								
I	0.033	0.019	0.012	0.015	0.031	0.018	0.015	0.018
II	0.057	0.034	0.031	0.032	0.119	0.058	0.034	0.021
IIIA	0.079	0.043	0.044	0.040	0.215	0.105	0.047	0.007
IIIB	0.079	0.050	0.042	0.025	0.108	0.030	0.032	0.007
2015								
I	0.035	0.023	0.017	0.027	0.132	0.082	0.082	0.009
II	0.045	0.034	0.036	0.040	0.263	0.110	0.074	0.013
IIIA	0.058	0.043	0.050	0.041	0.351	0.138	0.091	0.013
IIIB	0.056	0.032	0.048	0.024	0.305	0.025	0.084	0.013
Percentage change								
I	5	20	41	81	327	347	466	–47
II	–20	1	17	23	122	88	116	–37

Table 3.1—Continued

AFQT Category	Recruiting Rate				Non-HSDG Share			
	Army	Navy	Marine Corps	Air Force	Army	Navy	Marine Corps	Air Force
IIIA	–26	1	13	0	63	32	92	84
IIIB	–29	–36	12	–3	182	–17	163	91

NOTE: Predicted from RMC/wage coefficients in Tables B.6 and B.7 in Appendix B. We set recruiting goal, deployment, and unemployment at their means and turned female and post-2009 indicators off. The RMC/wage ratio is for men. The RMC/wage ratio was 1.36 in 1999 and 2.08 in 2015. Shading indicates that the prediction was based on RMC/wage coefficients that were statistically significantly different from 0.

wage ratios. For the Navy, the statistically significant change is limited to the IIIB recruiting rate, which is more than one-third lower (36 percent). This suggests that the Navy responded to higher RMC/wage ratios by bringing in fewer recruits who were deemed not high quality. For the Marine Corps and Air Force, there were increases in the category I and II recruiting rates, suggesting greater emphasis on recruiting from the upper tail of the AFQT distribution among HSDGs. The change in the II recruiting rate, for instance, was 17 percent for the Marine Corps and 23 percent for the Air Force.

The predicted changes for the non-HSDG share of accessions are nearly all statistically significant, and most are positive, except for the Navy IIIB and Air Force categories I and II. The Army, Navy, and Marine Corps, in contrast, showed a large percentage change in the non-HSDG share in the top categories (I and II). This suggests that these services responded to the increase in the RMC/wage ratio by bringing in more high-scoring non-HSDGs, yet the overall picture is mixed. The Army and Marine Corps also brought in more IIIAs and IIIBs, as did the Air Force.

The role of the Post-9/11 GI Bill is important in explaining the association between the RMC/wage ratio and the non-HSDG share shown in Table 3.2 and the differences in results between Tables 3.1 and 3.2. Specifically, the large percentage change in non-HSDG share seen in Table 3.2 is often larger than that in Table 3.1. Much of this disparity comes from the post-2009 indicator being off in the Table 3.2 predictions, while, for Table 3.1, it was off in 1999 and on in 2015. When we redid the percentage changes in Table 3.2 with the post-2009 indicator off in 1999 but on in 2015 (Table 3.3), this change alone resulted in much smaller percentage changes that are more in keeping with those in Table 3.1. This points to the apparent importance of the introduction of the Post-9/11 GI Bill in August 2009. Comparing Tables 3.2 and 3.3, we see that these changes tamp down the increase in non-HSDG share associated with the RMC/wage ratio and, in some cases, change an increase to a decrease (e.g., Army II, IIIA, and IIIB have increases in Table 3.2 and decreases in Table 3.3). Thus, bringing in the post-2009 indicator results in changes that are far more similar to the predictions in Table 3.1.

Sensitivity Tests and Limitations

We estimated different versions of the models to see whether the findings were robust. Given the relative smallness of the samples, we were not surprised to find some change in the point estimates of the coefficients, yet the pattern of results described above remained. The variants we tried were estimating the models separately by gender (i.e., the same models but with the data and estimation split by gender), excluding recruiting goal and deployment, excluding post-2009 and an interaction of post-2009 with IIIB, and excluding unemployment.

Our data are highly aggregate and offer limited variation, which might have affected the estimates. Unemployment coefficients were expected to be positive in the recruiting-rate regression and negative in the non-HSDG regression, but the signs are mixed. Studies using time-series and cross-section data (disaggregated across recruiting areas) find positive effects of unemployment (see, e.g., Warner, Simon, and Payne, 2003). On the other hand, the results for the post-2009 indicator are consistent with expectations. Army lost its edge from the Army College Fund. We found that its IIIB recruiting rate of high school graduates increased after 2009. This occurred as its HSDG share increased. The other services decreased their IIIB recruiting rates (took fewer lower-scoring high school graduates), and two services, the Navy and Marine Corps, decreased their non-HSDG shares of accessions.

Conclusion

The analysis in this chapter addresses the third question asked for this report—namely, was the increase in military pay since 1999 associated with an increase in the quality of recruits? At the beginning of this chapter, we showed that a measure of pay relevant to recruiting, the RMC/wage ratio for an E-4 with four YOSs, rose significantly from 1999 to 2015. Figures depicting the increase in recruit quality during this period indicated that it varied by service. We used regression analysis to estimate the association between RMC and recruit quality while controlling for other factors.

Table 3.3
Predicted Recruiting Rate and Non–High School Diploma Graduate Share for Men, by Regular Military Compensation Percentile, Allowing Only the Regular Military Compensation/Wage Ratio and Post-2009 Indicator to Change, and Percentage Change, 1999 and 2015

AFQT Category	Recruiting Rate				Non-HSDG Share			
	Army	Navy	Marine Corps	Air Force	Army	Navy	Marine Corps	Air Force
1999								
I	0.033	0.019	0.012	0.015	0.031	0.018	0.015	0.018
II	0.057	0.034	0.031	0.032	0.119	0.058	0.034	0.021
IIIA	0.079	0.043	0.044	0.040	0.215	0.105	0.047	0.007
IIIB	0.079	0.050	0.042	0.025	0.108	0.030	0.032	0.007
2015								
I	0.037	0.026	0.017	0.026	0.036	0.036	0.033	0.013
II	0.048	0.039	0.035	0.038	0.080	0.048	0.029	0.018
IIIA	0.061	0.048	0.049	0.039	0.117	0.062	0.036	0.018
IIIB	0.082	0.017	0.041	0.004	0.051	0.013	0.017	0.017
Percentage change								
I	11	35	39	72	16	94	124	–27
II	–16	13	15	17	–32	–17	–15	–14

Table 3.1—Continued

AFQT Category	Recruiting Rate				Non-HSDG Share			
	Army	Navy	Marine Corps	Air Force	Army	Navy	Marine Corps	Air Force
IIIA	–23	13	11	–4	–46	–41	–23	152
IIIB	4	–66	–4	–86	–53	–58	–46	148

NOTE: Predicted from RMC/wage coefficients in Tables B.6 and B.7 in Appendix B. We set recruiting goal, deployment, and unemployment at their means and set female indicator to off. The post-2009 indicator was off for 1999 and on for 2015, and the RMC/wage ratio is for men. The RMC/wage ratio was 1.36 in 1999 and 2.08 in 2015. Shading indicates prediction is based on RMC/wage coefficients that are statistically significantly different from 0.

Broadly, the Navy, Marine Corps, and Air Force increased recruit quality as the RMC/wage ratio increased, but the Army did not. The Marine Corps and Air Force increased their category I and II recruiting rates, and the Navy strongly decreased its IIIB recruiting rate. In contrast, the association between the RMC/wage ratio and the Army recruiting rate for II, IIIA, and IIIB was negative.

With respect to the non-HSDG share of accessions, the increase in the RMC/wage ratio was associated with an increase in the share for categories I and II for the Army, Navy, and Marine Corps. That is, the share for individuals scoring high on the AFQT but not graduating from high school was positively associated with the RMC/wage ratio. This was also true for the shares for IIIA and IIIB for the Army and Marine Corps. Air Force behavior was different, with increases in the non-HSDG shares for IIIA and IIIB but, unexpectedly, decreases for I and II, as the RMC/wage ratio rose. Even so, the Air Force share of non-HSDG remained extremely low, at about 2 percent.

Our regression models are reduced form and did not identify the causal effect that military or civilian pay has on recruiting outcomes, although this does not mean that RMC/wage effects are absent. We discuss other factors that can affect recruiting outcomes, including recruiting goal, deployment, unemployment, and the Post-9/11 GI Bill. We have suggested that these variables, along with the RMC/wage ratio, are beyond the immediate influence of the services, although we recognize that there could be lagged time-series effects that induce higher military pay when prior-period recruiting is difficult. We have also argued that decisions about recruiting resources—recruiters, advertising, bonus availability and amount, recruiting stations, eligibility policy, and so forth—depend on the explanatory variables we have included. With fuller data and more observations, we would choose to keep the variables we have and add the recruiting resource variables. Ideally, each recruiting resource could be varied experimentally in order to avoid endogeneity, but the importance of meeting recruiting missions has often taken precedence over experimentation. The method of instrumental variables can control for endogeneity, but only if there are instruments (other variables) correlated with the endogenous variable, uncorrelated with the recruiting outcome errors, and not thought to enter the structural equation. Unemployment is a possible instrument for bonuses, for instance, but it also affects the supply of recruits and should enter the structural recruiting outcome equations.

How might one interpret the RMC/wage coefficients in the reduced-form regressions? Holding constant recruiting goal, deployment, and unemployment, an increase in the RMC/wage ratio can be expected to increase the supply of recruits of all qualities and enable recruiters to be more successful in recruiting high-quality prospects—a positive, direct structural effect. At the same time, a higher RMC/wage ratio can be expected to decrease the demand for recruiting resources (e.g., recruiters, bonuses, advertising) needed to meet the goal—a negative, indirect effect. The possible cutback in recruiting resources implies that the reduced-form coefficients on the RMC/wage ratio might underestimate the ratio's causal effect on recruiting outcomes. If a poor recruiting outcome in the previous period led to a higher level of military pay in the current period, other things equal, this too would tend to bias the RMC/wage coefficient down from its true effect on recruit supply.

The findings in this chapter raise further questions: Is today's high level of recruit quality the appropriate level, and could it have been attained at lower cost? In Chapter Four, we discuss these questions, but an analysis of them was beyond the scope of this research.

CHAPTER FOUR

Closing Thoughts

Our findings are relevant to the following questions:

- How does military pay for active-component personnel compare to civilian pay?
- Has the position of military pay improved or worsened since 2009, when the 11th QRMC last benchmarked military pay?
- Given that military pay has increased since 1999, when the 9th QRMC first benchmarked military pay, was the increase in military pay since 1999 associated with an increase in recruit quality?

In this chapter, we summarize our findings on the military/civilian pay comparisons, discuss why the recruiting-rate results for the Army differed from those for the other services, ask whether recruit quality is at the right level today and whether the increase in RMC was cost-effective, and raise questions for additional research.

Findings in Brief

At What Percentile of Civilian Pay Did Regular Military Compensation Stand in 2016?

We found that RMC in 2016 was at the 84th percentile of comparable civilians' pay for active-component enlisted personnel and at the 77th percentile for active-component officers. As discussed, the 9th QRMC noted that many enlisted members have some college and recommended that military pay be at around the 70th percentile for that level of education. In 2016, we found, RMC was at the 87th percentile for enlisted with some college and the 93rd percentile for those with high school. For officers, RMC was at the 86th percentile for officers with bachelor's degrees and the 70th percentile for officers with master's degrees or higher. We also compared RMC to civilian wages over time, from 2000 to 2016, for selected age/education groups. These comparisons showed that RMC increased steadily, more than civilian pay did, from 2000 to 2010 and leveled off afterward.

At What Percentile of Civilian Pay Did Regular Military Compensation Stand in 2009?

RMC in 2009 was at the 84th percentile of pay for civilians comparable to enlisted members and at the 77th percentile of pay for civilians comparable to officers—the same as 2016. RMC was at the 85th percentile for enlisted with some college and the 91st percentile for those with high school. RMC was at the 87th percentile for officers with bachelor's degrees and the 69th percentile for those with master's degrees or higher. Our RMC percentiles for 2009 are somewhat below those of the 11th QRMC estimates—90th percentile for enlisted personnel and 83rd for officers—and we attribute the difference to methodological reasons described in Chapter Two.

Our finding that the RMC percentile was nearly the same in 2016 as 2009 might be surprising because there were years when basic-pay raises were below the ECI (2014 through 2016), although, in our view, the ECI is not a reliable guide for military and civilian pay comparisons for various reasons, including that it is not adjusted for the military education distribution.

How Did Recruit Quality Change as Regular Military Compensation Rose?

We used regression models to isolate the relationship between the RMC/wage ratio and recruiting outcomes. We found that, as the RMC/wage ratio increased, recruit quality increased in the Navy, Marine Corps, and Air Force but not in the Army.

Why Were the Army Results Different?

We do not have a definitive answer but can suggest some possibilities. One is that, for reasons not controlled in our models, Army recruiting became more difficult than other services' recruiting. Under this hypothesis, the supply curve of recruits shifted back, decreasing the number of recruits available at each wage. Assuming that the supply curve was upward sloping (i.e., the number of recruits available increased as the wage increased), the increase in RMC from 1999 to 2015 would have helped to offset the decrease and therefore prevent it from being worse than it otherwise would have been. Still, there was a net *negative* association between the RMC/wage ratio and Army recruiting rates for categories II, IIIA, and IIIB. A shift back in supply might have reflected differences between youth attitudes toward the Army and those toward the other services, which is consistent with this hypothesis. Data from the DoD Youth Polls from between 2001 and 2015 show a downward trend in youths' propensity to enlist, but the trend was quite similar across the services.[1] A related explanation is that deployments in support of operations in Iraq and Afghanistan had a negative effect on recruiting beyond that controlled in our models. We estimated small deployment coefficients in our Army models, and we have suggested that the Army's more-extensive use of bonuses and higher bonus amounts tended to nullify a negative effect of deployment. But perhaps the deployment variable and the implicit bonus response to high deployment does not fully control for the impact of deployment, and the uncontrolled and unobserved aspect of deployment might have been positively correlated with RMC and had a negative effect on recruiting.

Another possibility is that the Army set quality goals and programmed recruiting resources to sustain, but not increase, accession quality. Army accession quality has been above the DoD guidance of 60 percent for the percentage of accessions scoring in the upper half of the AFQT, although this percentage has declined over time. Also, in some years, the Army was below the guidance of 90 percent for HSDGs

---

[1] The DoD Youth Polls began in April 2001 (Office of People Analytics, 2018).

(but is well above this guidance today). A policy of not increasing accession quality would be appropriate if the incremental benefit relative to the incremental cost of HSDG II–IIIB recruits were lower than that for category I–IIIB non-HSDG recruits. This explanation is consistent with the *positive* association between the RMC/wage ratio and the share of accessions who were non-HSDGs in categories I through IIIB (Figure 3.10 in Chapter Three). This might be especially relevant to Army expansion during the 2000s. The recruiting goal went from 73,800 in 2003 to 80,000 in 2005 and remained there for several years (Nataraj et al., 2017). The greater emphasis on recruiting high-scoring non-HSDGs would have helped the Army meet its recruiting targets while trying to sustain its AFQT mix. At the same time, the Army explored nontraditional measures of quality, including the ARMS and TTAS programs noted in Chapter Three, which widened recruit selection criteria and expanded the pool of recruitable youths.

Other factors might also be relevant. Higher RMC might have affected recruiter effort. Research suggests that recruiters exert less effort when pay and other recruiting resources are plentiful (Dertouzos, 1985). Maybe Army recruiters reduced their effort on II–IIIB HSDGs when RMC rose. The Army faced difficult recruiting conditions in the mid-2000s and greatly increased bonus usage and amounts, which is contrary to that explanation (Hosek and Martorell, 2009), although RMC had also increased. RMC continued to rise after the mid-2000s, and the Great Recession took hold in 2008, so Army recruiters might have had some possibility of reducing effort in the recession era. But during this period, the Army's percentage of accessions who were HSDGs rose rapidly, suggesting continued recruiter effort. Still, the percentage of accessions in categories I through IIIA continued to drift downward, as did the percentage of I–IIIA HSDGs.

We explored some of these issues theoretically in a two-goal model of recruiting resource allocation (Appendix D). In the model, a service has goals for high-quality and non–high-quality recruits, and the goals sum to the overall quantity goal. There are production functions for high-quality and non–high-quality recruits. The model recognizes that each service might face its own recruiting markets and have its own recruit production functions. The services set their own quality goals to be consistent with DoD guidance. Each production function has inputs specific to it that can be varied independently of inputs in the other production functions; these inputs are recruiter time and bonus. Each production function also has inputs common to both functions. These are advertising and waiver policy, which are under the service's control, and relative pay, deployment, and unemployment, not under its control. The level of a common input is the same in both production functions, but its effect on output can differ between the production functions, as can the effects of recruiter time and bonus. The model showed that an increase in RMC would lead to a decrease in recruiting resources, such as recruiter effort and bonuses, subject to meeting the quality and quantity goals and holding other variables constant (e.g., the unemployment rate). In other words, the model rationalizes a possible trade-off between RMC and a service's recruiting resources. The model also showed the optimal resourcing conditions subject to meeting the quantity goal and exceeding the quality goal.

Is Recruit Quality at the Right Level Today?

It is hard to know the right balance between recruit quality and cost. Defense capability gained from recruiting and retaining more high-quality recruits must be weighed against the added cost of higher RMC, which increases the entire personnel budget. Our analysis has shown where and to what extent recruit quality increased as military pay increased,[2] but it does not place a value on the increased quality. Valuing quality is the services' domain. In the absence of empirical evidence to the contrary, it might have been optimal for the Navy, Air Force, and Marine Corps to increase their percentages of high-quality recruits as RMC increased and for the Army to decrease its percentage.

As mentioned in the introduction, research has shown that the recruit quality a service brings in is approximately the quality it retains

---

[2] We found that RMC for an E-4 with four YOSs reached the 70th percentile of pay for high school graduates ages 23 to 27 shortly after the pay increases mandated by the NDAA for FY 2000. RMC was at the 69th percentile in 2001 and at the 74th percentile in 2002. By 2010, RMC was around the 85th percentile.

throughout that service member's enlisted career. Yet, other changes have occurred in the defense environment, and the 9th QRMC's 70th-percentile guidance might not be right for today. Changes in defense threats, readiness requirements, and military technology have shifted manpower requirements toward higher-AFQT recruits in some services. For example, the Navy's weapon systems have become more complex in the past decade. Its widened use of software-based technology is needed to support network-centric warfare, causing an increase in the demand for personnel with information technology skills and the ability to handle complex information in decisionmaking (Wenger, Miller, and Sayala, 2010).

Was the Increase in Regular Military Compensation Cost-Effective?

The higher-than-ECI increases in basic pay until 2010 resulting from the NDAA for FY 2000 and subsequent legislation helped to sustain recruiting and retention during wartime, when frequent, long deployments to Iraq and Afghanistan stressed recruiting and retention (Asch, Heaton, et al., 2010; Hosek and Martorell, 2009). DoD-wide, recruit quality has increased considerably since the early 2000s. The increases in basic pay, BAH, and bonuses no doubt helped the services meet their manning requirements during this period. Since 2009, we found, the RMC percentile has remained constant at its new, higher level.

It is clear that military pay is an important force-management tool and was critical in the 2000s to sustain the force during wartime. But the increase in recruit quality for three of the four services between 1999 and 2015 raises the question of whether achieving that quality could have been accomplished in a more cost-effective manner.

Although estimates differ on the marginal cost of pay and recruiting resources, researchers in virtually every relevant study have found that military pay is the costliest approach for enlisting high-quality recruits (Orvis, Garber, et al., 2016; Asch, Heaton, et al., 2010; Simon and Warner, 2007). RMC is a blunt instrument that is not targeted to occupational specialties for which recruiting or retention shortfalls occur. An increase in RMC affects the cost of personnel budgets in every service, while an increase in a service's recruiting resources, such as recruiters, enlistment bonuses, advertising, and recruiting stations and equipment, is specific to its recruiting budget, and some of those resources, such as bonuses, can be targeted.

If the RMC percentiles were decreased from their current level, one possibility is that Navy, Marine Corps, and Air Force would allow recruit quality to decline. But if today's quality is needed for today's manning requirements, these services would need to increase their recruiting resources to make up for the decrease in RMC percentiles. The Army's response would be similar, but, more than with the other services, it would be constrained from below by DoD's quality floor for categories I through IIIA.

Additional Research Questions

The 9th QRMC drew attention to the increase in education as grade and YOSs increased and presented evidence that this was a trend that had begun in the 1980s and could be expected to continue. The data for 2016 indicate that the trend has indeed continued. This means not only that active-component members are interested in adding to their education while in service but also that the services have been able to provide the incentive, opportunity, and support for doing so. Given the importance of educational opportunity, we suggest possible future research that would focus on in-service education:

- What is the incentive to obtain more education? The answer might lie in education being a factor in determining promotion and in creating better civilian employment and earning opportunities.
- To what extent have the opportunities to pursue education increased over time? For instance, is this a positive externality in a military that has become more internetted, making online courses an option for personnel in remote locations, and how important

have efforts been to facilitate the transfer of course credits when a member changes duty stations?
- Finally, is the added education on par with civilian education—for instance, does an associate's degree obtained in the military bring civilian wages on par with what a civilian associate's degree does?

APPENDIX A

Regular Military Compensation Percentile and Regular Military Compensation/Wage Ratio

This appendix describes how we smoothed the RMC percentile for an E-4 with four YOSs relative to the civilian-wage distribution for high school graduates ages 18 to 22 who worked more than 35 weeks in the year and had more than 35 usual hours of work. The smoothed values are shown in Figure 3.4 in Chapter Three. This appendix also describes the ratio of RMC to the median civilian wage. RMC is based on Greenbook data, and civilian wages come from March CPSs. The RMC/wage ratio is shown in Figure 3.5 in Chapter Three.

Smoothing the Regular Military Compensation Percentile

Raw RMC percentiles vary considerably from year to year. Some variation comes from annual increases in RMC resulting from increases in basic pay and BAH, but much of the variation comes from the smallness of CPS samples. For high school graduates ages 18 to 22, sample sizes for each year of data range from 150 to 250 observations for men and the same for women. These sample sizes do not provide dense enough coverage for a precise estimate of the wage distribution or the RMC percentile. We used a smoothing method to adjust for the variation.

There are different approaches to smoothing. One approach is a kernel density estimator to smooth the wage distribution each year. But that approach does not use data from adjacent years and cannot be relied on to provide year-to-year continuity. Instead, we estimated a log wage model to identify the mean and variance of the wage distribution, allowing for a common trend in real wages. The estimation used the tabulated wages at the 30th, 40th, 50th, 60th, 70th, 80th, and 90th percentiles. Then, using the estimated wage distribution parameters, we inferred the RMC percentile. The approach smooths the RMC percentile and provides year-to-year continuity in the wage distribution as real wages change over time.

Using Wage Percentiles to Estimate the Log Wage Distribution

Let p be the percentile (e.g., $p = 0.6$ at the 60th percentile), and let F be the standard normal distribution with mean 0 and standard deviation σ. Assume that the wage is log-normally distributed, so

$$p = F\left[\frac{\ln w_p - \mu}{\sigma}\right].$$

Taking the inverse normal, the log wage at percentile p is $\ln w_p = \mu + \sigma F^{-1}[p]$.

We tabulated CPS wage data to find wages at the 30th through 90th percentiles for each year, 1999 through 2015. These wages are the observations on $\ln w_p$. Thus, for a given group (e.g., 18- to 22-year-old high school graduates), the log wage at percentile p in year t is

$$\ln w_{pt} = \mu_t + \sigma F^{-1}[p].$$

At the 50th percentile, $F^{-1}[0.5] = 0$, and it follows that the mean of the log wage distribution, μ_t, equals the log of the wage at the 50th percentile. We computed the wage at the 50th percentile and used it to obtain an estimate of the mean of the log wage distribution: $\mu_t = \ln w_{0.5t}$ at each year and, in particular, for the base year of our data, 1999. We refer to 1999 as period 0.

For small changes, we approximated the year-to-year wage change as a percentage change from the wage in period 0:

$$w_{pt} = e^{\mu_0 + \sigma F^{-1}[p]} e^{\delta t}.$$

Taking logs, we have

$$\ln w_{pt} = \mu_0 + \sigma F^{-1}[p] + \delta t.$$

We can replace μ_0 with ln $w_{0.5t}$ + ε_{pt} given that $w_{0.50}$ is computed from the data and its log is an estimate of the mean. Subtract ln $w_{0.50}$ from both sides to obtain ln w_{pt} – ln $w_{0.50}$ = $\sigma F^{-1}[p] + \delta t + \varepsilon_{pt}$. In the usual regression format, this can be thought of as ln w_{pt} – ln $w_{0.50}$ = $\beta_1 F^{-1}[p] + \beta_2 t + \varepsilon_{pt}$, where β_1 is an estimate of σ, the standard deviation of the log wage; β_2 is an estimate of δ, the annual percentage change in the wage; and there is no intercept. Values for $F^{-1}[p]$ at each percentile came from the inverse normal function evaluated at the given percentile. The variable t is the year. This approach assumes that the standard deviation does not change during the observation period, and the mean evolves according to the time trend. For each group, there are wages for seven percentiles in each of 17 years, 1999 through 2015, for a total of 119 observations. There are two parameters to estimate.

Using the estimated parameters, the predicted wage in period t at percentile p is

$$w_{pt} = e^{\ln w_{0.50} + \hat{\sigma} F^{-1}[p] + \hat{\delta} t}.$$

Also, for a given value of the wage, the corresponding percentile is derived as follows:

$$\ln w_t - \ln w_{0.50} = \hat{\sigma} F^{-1}[p] + \hat{\delta} t$$

$$F^{-1}[p] = \frac{\left(\ln w_t - \ln w_{0.50} - \hat{\delta} t\right)}{\hat{\sigma}}$$

$$p = F\left[\frac{\ln w_t - \ln w_{0.50} - \hat{\delta} t}{\hat{\sigma}}\right].$$

Letting RMC_t stand in for wage, its percentile is

$$p = F\left[\frac{\ln RMC_t - \ln w_{0.50} - \hat{\delta} t}{\hat{\sigma}}\right].$$

Parameter Estimates and Goodness of Fit

Table A.1 reports the regression results. The standard deviation of the log wage distribution is 0.457 for men and 0.402 for women, both highly significant. Also, for interpretability, the estimates for δ are reported as the annual percentage change (i.e., as 100 times the estimated coefficient). For example, the δ estimate of –0.577 for male high school graduates reported in the table means that wages trended down by 0.577 percent per year from 1999 to 2015. Similarly, the reported standard error of δ is 100 times the estimated standard error. The wage data are in 2014 dollars. The trend estimate is statistically significant for men and women. The models fit the data well, with an R^2 of 0.93 for men and 0.89 for women.

Table A.1
Regression Results for 18- to 22-Year-Old High School Graduates

Group	Coefficient	Standard Error	t	P>|t|
Male				
Standard deviation of log wage distribution	0.457	0.012	37.68	0.000
Time trend	–0.577	0.085	–6.78	0.000
R^2	0.93			
Female				
Standard deviation of log wage distribution	0.402	0.013	30.79	0.000
Time trend	–1.160	0.092	–12.68	0.229
R^2	0.89			

Raw and Smoothed Regular Military Compensation Percentiles

Table A.2 contains the raw and smoothed (predicted) RMC percentiles. In making predictions, we used the coefficients in Table A.1 and

Table A.2
Regular Military Compensation Percentiles for High School Graduates, Ages 18 to 22, as Percentages

	Men		Women	
Calendar Year	Raw	Smoothed	Raw	Smoothed
1999	79	75	94	86
2000	76	77	88	87
2001	81	79	86	90
2002	80	83	90	92
2003	89	86	96	94
2004	88	85	91	94
2005	88	87	95	95
2006	87	88	91	96
2007	92	89	99	97
2008	94	89	91	97
2009	90	91	94	98
2010	96	92	100	98
2011	94	91	100	98
2012	89	92	100	98
2013	82	93	88	98
2014	86	93	98	99
2015	90	94	99	99

SOURCES: RMC data are from Directorate of Compensation, 1999–2015. We computed civilian wages from March CPS data for these years (U.S. Census Bureau, 2018).

NOTE: RMC is for an E-4 with four YOSs. Civilian wages are for 18- to 22-year-old workers with high school who worked more than 35 weeks in the year and had more than 35 usual weekly hours of work.

the median weekly wage for 1999, which was $546.54 for men and $483.13 for women.

The Regular Military Compensation/Wage Ratio

We again used RMC for an E-4 with four YOSs and wages for civilians ages 18 to 22 and who worked more than 35 weeks in the year and had usual weekly hours of more than 35. The wage ratio is RMC divided by the median wage (the wage at the 50th percentile of the wage distribution). Table A.3 shows the raw ratio, as well as the ratio predicted by fitting a line to the raw values. In our regression analysis, we used the raw values.

Table A.3
Regular Military Compensation/Wage Ratio for High School Graduates, Ages 18 to 22

	Men		Women	
Calendar Year	**Raw**	**Predicted**	**Raw**	**Predicted**
1999	1.36	1.45	1.54	1.61
2000	1.38	1.49	1.68	1.66
2001	1.53	1.53	1.61	1.71
2002	1.57	1.57	1.71	1.76
2003	1.76	1.61	1.86	1.82
2004	1.70	1.65	2.09	1.87
2005	1.64	1.69	1.90	1.92
2006	1.76	1.72	1.93	1.97
2007	1.82	1.76	2.09	2.03
2008	1.72	1.80	2.04	2.08
2009	1.81	1.84	2.15	2.13
2010	2.02	1.88	2.31	2.18
2011	2.12	1.92	2.25	2.24
2012	2.03	1.96	2.22	2.29
2013	1.85	1.99	2.17	2.34
2014	1.82	2.03	2.55	2.39
2015	2.08	2.07	2.36	2.45

SOURCES: RMC is from Directorate of Compensation, 1999–2015. We computed civilian wages from March CPS data (U.S. Census Bureau, 2018).

NOTE: RMC is for an E-4 with four YOSs. The wage ratio is RMC divided by the median wage for 18- to 22-year-old workers with high school (and not additional) education who worked more than 35 weeks in the year and had more than 35 usual weekly hours of work.

APPENDIX B

Recruiting Rates for Armed Forces Qualification Test Categories I Through IIIB and Regression Estimates

Recruiting Rates

The recruiting rate is the ratio of NPS HSDG enlisted accessions to the population of high school completers, net those who went on to complete four or more years of college. Accession data are from the MEPS file. Data on high school completers and on the percentage of high school completers who had completed four or more years of college by ages 25 to 29 are from NCES. We calculated recruiting rates by AFQT category. The category is given directly in MEPS data, but it is not present in NCES data. To allocate our adjusted high school completer population by AFQT category, we used the 1997 National Longitudinal Survey of Youth (NLSY), which administered the AFQT to a representative sample of youths.[1]

NCES provides data on recent high school completers, by gender, for 1960 through 2014 (NCES, 2015). We also drew on NCES data to calculate the percentage of 25- to 29-year-olds who completed bachelor's degrees or higher conditional on completing high school or higher

[1] We explored an alternative adjustment—namely, the percentage enrolled in higher education in the October after high school completion. The number enrolled was higher than the number of eventual completers, so the adjustment would subtract more people from the number of high school completers. However, the remaining population had similar movement over time to the population adjusted for higher-degree completers, which is the approach we used. Therefore, if we had used the "enrolled" adjustment in estimating the "qualified military available" population, that adjustment would have provided similar recruiting-rate regression estimates except for a lower intercept. We preferred our approach because some of the enrolled population becomes available for recruiting even if they are not available at a point in time (i.e., in the October after completing high school).

(NCES, 2016). We assumed a modal age of 18 for high school completers and a modal age of 27 for the 25- to 29-year-olds who completed bachelor's degrees or higher (i.e., nine years later: age 27 minus age 18). The completion-rate data were available through 2016. We fitted linear trend models to the higher-degree (bachelor's or higher) completion data and used the estimated trend models to predict the higher-degree completion rates for high school completers for the span covered by our data, 1999 to 2015. We deducted these percentages from the population of completers. What remained was the number of high school completers not expected to later complete four-year degrees or higher. We assumed that this was the population to be recruited into the military.

The 1997 NLSY is the most recent renorming of the ASVAB (on which the AFQT score is based). The NLSY provides information on the AFQT score distribution for 18- to 23-year-old men and women. We used the percentage of the population in each category to estimate the percentage of our net high school completer population by category. This did not account for the possibility that the AFQT distribution conditional on high school completion differs from that of the unconditional population. The high school completion rate in 2000 was 87 percent for men and 89 percent for women (NCES, 2016), suggesting that the AFQT distribution for high school completers is likely to be close to that for the 18- to 23-year-old population overall.

Table B.1 shows the total number of male and female high school completers, the percentage of completers expected to complete four or more years of college, and the number of high school completers net of the latter. Tables B.2 through B.5 show the recruiting rates by service and gender for AFQT categories I, II, IIIA, and IIIB.

Regression Estimates

Tables B.6 through B.9 contain the regression results. The upper panel of each table contains the regression coefficient estimate, the R^2, and the number of observations, and the lower panel contains the pay ratio estimates for each AFQT category. The regressions interact the pay ratio with each AFQT category, with IIIA serving as the base group. The interactions allow the coefficient on the pay ratio to differ by AFQT category. These interactions enable us to provide estimates of how the relationship between the RMC/wage ratio and recruiting outcomes varies by AFQT category. For instance, the pay ratio coefficient for category I, shown in the lower panel of each table, is the sum of the pay ratio coefficient for IIIA and the "pay ratio × category I" coefficient in the upper panel, and similarly for the other categories.

Table B.1
High School Completers, 1999 Through 2014, Net Those Predicted to Complete Four or More Years of College

Calendar Year	High School Completers, in Thousands		Percentage Predicted to Complete Bachelor's or Higher		High School Completers Net of Predicted Bachelor's or Higher Completers, in Thousands	
	Male	Female	Male	Female	Male	Female
1999	1,474	1,423	30.5	37.5	1,024	890
2000	1,251	1,505	31.1	38.0	862	932
2001	1,277	1,273	31.7	38.6	872	781
2002	1,412	1,384	32.3	39.2	956	842
2003	1,306	1,372	32.8	39.8	877	826
2004	1,327	1,425	33.4	40.3	883	850
2005	1,262	1,414	34.0	40.9	832	835
2006	1,328	1,363	34.6	41.5	869	798
2007	1,511	1,444	35.2	42.1	979	837
2008	1,640	1,511	35.8	42.6	1,054	866
2009	1,407	1,531	36.4	43.2	895	869
2010	1,679	1,482	36.9	43.8	1,059	833
2011	1,611	1,468	37.5	44.4	1,006	817

Table B.1—Continued

Calendar Year	High School Completers, in Thousands		Percentage Predicted to Complete Bachelor's or Higher		High School Completers Net of Predicted Bachelor's or Higher Completers, in Thousands	
	Male	Female	Male	Female	Male	Female
2012	1,622	1,581	38.1	44.9	1,004	870
2013	1,524	1,453	38.7	45.5	935	792
2014	1,423	1,445	39.3	46.1	864	779
2015[a]	1,580	1,502	39.9	46.7	950	801

SOURCE: The numbers of high school completers are from NCES, 2015. The percentages predicted to complete bachelor's degrees or higher, conditional on high school completion, are based on raw data from NCES, 2016, on the percentage of 25- to 29-year-olds who completed high school or higher and the percentage who completed bachelor's degrees or higher.

NOTE: We fitted a linear trend to the percentage of 25- to 29-year-olds who completed bachelor's degrees or higher, conditional on completing high school or more, for 2005 through 2016. We assumed a median age of 27 for the 25- to 29-year-olds and a median age of 18 for high school completers, a nine-year difference, then used the linear trend to predict the percentage of high school completers in 1999 through 2015 who would complete bachelor's degrees or higher by nine years later.

[a] The number of high school completers in 2015 was predicted from a linear trend model fitted to high school completers in 1999 through 2014.

Table B.2
Army Recruiting Rates: Armed Forces Qualification Test Categories I, II, IIIA, and IIIB, as Percentages

	I		II		IIIA		IIIB	
Year	Men	Women	Men	Women	Men	Women	Men	Women
1999	2.15	0.4	4.15	1.3	6.55	2.4	7.61	2.1
2000	2.68	0.4	5.29	1.3	7.98	2.2	8.25	2.6
2001	2.74	0.5	5.23	1.5	7.90	2.6	7.77	3.0
2002	3.51	0.5	5.27	1.4	7.03	2.2	6.52	2.3
2003	4.13	0.6	5.68	1.3	7.47	2.1	6.33	2.3
2004	5.44	0.7	7.09	1.5	9.25	2.3	7.78	2.3
2005	4.71	0.6	5.85	1.2	7.45	1.9	6.91	2.1
2006	3.95	0.5	4.98	1.1	5.81	1.5	6.52	2.0
2007	3.09	0.4	3.88	1.0	4.51	1.2	4.99	1.6
2008	3.08	0.4	4.05	1.0	4.91	1.3	5.68	1.9
2009	4.07	0.4	5.14	1.0	6.35	1.3	7.68	1.9
2010	4.36	0.6	5.50	1.1	6.72	1.7	8.49	2.5
2011	4.11	0.5	5.09	1.0	6.17	1.5	8.16	2.5
2012	3.36	0.4	4.91	0.9	6.46	1.3	7.95	2.0
2013	3.79	0.5	5.81	1.1	8.11	1.9	10.44	2.9
2014	3.06	0.4	4.85	0.9	7.10	1.7	8.72	2.3
2015	2.87	0.4	4.52	0.9	6.55	1.7	8.68	2.6

Table B.3
Navy Recruiting Rates: Armed Forces Qualification Test Categories I, II, IIIA, and IIIB, as Percentages

	I		II		IIIA		IIIB	
Year	Men	Women	Men	Women	Men	Women	Men	Women
1999	2.0	0.3	3.7	1.0	4.8	1.5	6.2	1.8
2000	2.2	0.3	4.3	0.9	5.6	1.5	7.5	1.7
2001	2.0	0.3	4.0	1.1	5.3	1.8	7.5	2.0
2002	2.1	0.3	3.6	0.9	4.5	1.4	5.8	1.4
2003	2.3	0.3	3.5	0.8	4.4	1.4	5.5	1.1
2004	3.1	0.4	4.4	1.0	5.1	1.4	5.5	1.1
2005	2.8	0.4	4.3	0.9	5.9	1.4	6.5	1.3
2006	2.6	0.3	4.0	0.9	4.8	1.4	3.9	1.3
2007	2.1	0.3	3.1	0.8	3.8	1.1	3.7	1.2
2008	1.4	0.2	2.7	0.7	3.4	1.1	3.1	1.1
2009	1.9	0.3	3.2	0.8	3.8	1.2	2.9	0.9
2010	2.0	0.3	2.9	0.9	3.4	1.5	1.8	0.7
2011	2.2	0.4	3.2	1.0	3.8	1.6	1.1	0.6
2012	2.5	0.4	3.5	1.1	4.3	1.7	1.0	0.6
2013	2.7	0.4	3.9	1.2	4.8	1.9	2.0	1.0
2014	2.4	0.4	3.6	1.0	4.6	1.8	1.3	0.7
2015	2.1	0.4	3.4	1.1	4.4	2.0	1.0	0.8

Table B.4
Marine Corps Recruiting Rates: Armed Forces Qualification Test Categories I, II, IIIA, and IIIB, as Percentages

	I		II		IIIA		IIIB	
Year	Men	Women	Men	Women	Men	Women	Men	Women
1999	1.0	0.1	2.9	0.3	4.3	0.5	4.8	0.4
2000	1.1	0.1	3.1	0.2	4.7	0.4	5.2	0.3
2001	1.1	0.1	3.1	0.3	4.8	0.5	5.2	0.4
2002	1.2	0.1	3.1	0.3	4.6	0.5	4.6	0.3
2003	1.5	0.1	3.5	0.3	4.7	0.5	4.5	0.3
2004	1.9	0.1	3.8	0.3	5.2	0.5	4.8	0.3
2005	2.0	0.1	4.2	0.3	5.7	0.5	5.8	0.5
2006	1.8	0.1	3.6	0.3	4.7	0.4	4.8	0.5
2007	1.7	0.1	3.4	0.3	4.3	0.4	4.6	0.5
2008	1.6	0.1	3.4	0.3	4.5	0.4	4.6	0.4
2009	1.7	0.1	3.5	0.3	4.6	0.5	4.4	0.4
2010	1.5	0.1	2.9	0.3	3.8	0.5	3.1	0.4
2011	1.6	0.1	3.3	0.3	4.4	0.5	3.3	0.4
2012	1.6	0.1	3.4	0.3	4.7	0.5	3.3	0.4
2013	1.7	0.2	3.7	0.4	5.2	0.6	3.9	0.5
2014	1.4	0.1	3.1	0.4	4.6	0.7	3.4	0.5
2015	1.4	0.1	3.2	0.4	4.9	0.7	3.6	0.4

Table B.5
Air Force Recruiting Rates: Armed Forces Qualification Test Categories I, II, IIIA, and IIIB, as Percentages

	I		II		IIIA		IIIB	
Year	Men	Women	Men	Women	Men	Women	Men	Women
1999	1.3	0.3	3.0	1.1	3.7	1.8	2.3	1.3
2000	1.5	0.2	3.6	1.0	4.7	1.7	3.5	1.4
2001	1.6	0.3	3.7	1.2	4.6	1.9	3.2	1.4
2002	1.8	0.3	3.7	1.2	4.6	1.9	3.0	1.3
2003	2.2	0.4	4.1	1.2	4.7	1.9	2.5	1.0
2004	2.9	0.5	4.7	1.4	5.4	2.0	2.6	1.0
2005	1.8	0.3	2.7	0.8	3.3	1.1	1.7	0.6
2006	2.3	0.4	3.8	1.1	4.2	1.6	2.5	1.2
2007	1.8	0.3	3.1	1.0	3.3	1.4	2.1	1.0
2008	1.8	0.3	2.9	0.9	3.1	1.3	1.9	0.9
2009	2.7	0.4	4.0	1.0	4.1	1.4	2.3	0.9
2010	2.4	0.4	3.4	1.0	3.4	1.2	0.9	0.4
2011	2.5	0.4	3.9	1.1	4.2	1.5	0.1	0.1
2012	2.6	0.4	3.9	1.0	4.2	1.4	0.2	0.1
2013	2.4	0.4	3.8	1.1	4.0	1.3	0.2	0.1
2014	2.4	0.4	3.7	1.0	3.8	1.4	0.5	0.2
2015	2.2	0.3	3.2	1.0	3.2	1.3	0.7	0.3

Table B.6
Logit Regression of Recruiting Rate for AFQT Categories I Through IIIB, Army and Navy

	Army			Navy		
Variable	Coefficient	SE	*t*	Coefficient	SE	*t*
Pay ratio	–0.4567	0.1873	–2.44	0.0148	0.1926	0.08
Pay ratio × category I	0.5300	0.1756	3.02	0.2435	0.1274	1.91
Pay ratio × category II	0.1231	0.1381	0.89	–0.0002	0.1049	0.00
Pay ratio × category IIIB	–0.0535	0.2093	–0.26	–0.6598	0.2676	–2.47
Recruiting goal	0.0047	0.0026	1.81	0.0077	0.0061	1.27
Deployment	–0.0049	0.0067	–0.74	–0.0861	0.0632	–1.36
Unemployment	0.0231	0.0102	2.26	–0.0189	0.0094	–2.01
Category I	–1.6242	0.3407	–4.77	–1.1435	0.2330	–4.91
Category II	–0.5183	0.2637	–1.97	–0.2334	0.1880	–1.24
Category IIIB	0.0761	0.3383	0.23	1.0740	0.4456	2.41
Female	–1.2691	0.0773	–16.41	–1.1269	0.0639	–17.64
Female × category I	–0.7664	0.0938	–8.17	–0.8699	0.0798	–10.90
Female × category II	–0.1924	0.0829	–2.32	–0.2390	0.0683	–3.50
Female × category IIIB	0.1589	0.0987	1.61	0.1988	0.1320	1.51
Post-2009	0.0531	0.0716	0.74	0.1189	0.0479	2.48
Post-2009 × category IIIB	0.3547	0.0992	3.58	–0.7716	0.1575	–4.90
Constant	–2.2968	0.4124	–5.57	–3.2382	0.5789	–5.59
R^2	0.97			0.96		
Observations	136			136		
Pay ratio I	0.0733	0.2080	0.35	0.2583	0.1909	1.35
Pay ratio II	–0.3336	0.1656	–2.02	0.0146	0.1779	0.08
Pay ratio IIIA	–0.4567	0.1873	–2.44	0.0148	0.1926	0.08
Pay ratio IIIB	–0.5102	0.1752	–2.91	–0.6450	0.3304	–1.95

NOTE: Critical values for the *t*-statistic are 10 percent, 1.645; 5 percent, 1.96; and 1 percent, 2.58. SE = standard error.

Table B.7
Logit Regression of Recruiting Rate for AFQT Categories I Through IIIB, Marine Corps and Air Force

	Marine Corps			Air Force		
Variable	Coefficient	SE	*t*	Coefficient	SE	*t*
Pay ratio	0.1788	0.1259	1.42	0.0027	0.1203	0.02
Pay ratio × category I	0.3050	0.1275	2.39	0.8406	0.1556	5.40
Pay ratio × category II	0.0438	0.1023	0.43	0.2954	0.1052	2.81
Pay ratio × category IIIB	–0.0107	0.1967	–0.05	–0.0406	0.2774	–0.15
Recruiting goal	–0.0170	0.0057	–3.00	0.0242	0.0035	6.94
Deployment	0.0133	0.0069	1.93	–0.0765	0.0869	–0.88
Unemployment	–0.0158	0.0074	–2.13	–0.0054	0.0150	–0.36
Category I	–1.7184	0.2315	–7.42	–2.1512	0.2910	–7.39
Category II	–0.4261	0.1792	–2.38	–0.6347	0.1934	–3.28
Category IIIB	–0.0220	0.3389	–0.06	–0.4559	0.4608	–0.99
Female	–2.3234	0.0442	–52.52	–0.9970	0.0456	–21.85
Female × category I	–0.3875	0.0631	–6.14	–1.0105	0.0673	–15.02
Female × category II	–0.1141	0.0539	–2.12	–0.3269	0.0530	–6.16
Female × category IIIB	–0.1150	0.0816	–1.41	0.1108	0.1853	0.60
Post-2009	–0.0160	0.0448	–0.36	–0.0482	0.0577	–0.83
Post-2009 × category IIIB	–0.1494	0.0939	–1.59	–1.8932	0.2383	–7.94
Constant	–2.7497	0.3020	–9.10	–3.7742	0.2144	–17.61
R^2	0.99			0.94		
Observations	136			136		
Pay ratio I	0.4838	0.1368	3.54	0.8433	0.1836	4.59
Pay ratio II	0.2226	0.1038	2.14	0.2981	0.1313	2.27
Pay ratio IIIA	0.1788	0.1259	1.42	0.0027	0.1203	0.02
Pay ratio IIIB	0.1681	0.1924	0.87	–0.0379	0.2905	–0.13

NOTE: Critical values for the *t*-statistic are 10 percent, 1.645; 5 percent, 1.96; and 1 percent, 2.58. SE = standard error.

Table B.8
Logit Regression of Share of Non–High School Diploma Graduate Accessions for Armed Forces Qualification Test Categories I Through IIIB, Army and Navy

	Army			Navy		
Variable	Coefficient	Standard Error	*t*	Coefficient	Standard Error	*t*
Pay ratio	0.9505	0.3269	2.91	0.4349	0.4274	1.02
Pay ratio × category I	1.2273	0.2891	4.24	1.7487	0.3985	4.39
Pay ratio × category II	0.4079	0.2086	1.96	0.5267	0.3338	1.58
Pay ratio × category IIIB	0.8433	0.6708	1.26	−0.7071	0.6576	−1.08
Recruiting goal	0.0093	0.0090	1.04	0.0601	0.0097	6.19
Deployment	0.0161	0.0199	0.81	−0.3641	0.1564	−2.33
Unemployment	−0.1077	0.0291	−3.70	−0.0045	0.0244	−0.18
Category I	−3.8217	0.4985	−7.67	−4.2196	0.6691	−6.31
Category II	−1.2675	0.3673	−3.45	−1.3601	0.5501	−2.47
Category IIIB	−1.9623	1.0511	−1.87	−0.3777	1.0911	−0.35
Female	−0.9673	0.1186	−8.15	−0.7304	0.1717	−4.25
Female × category I	0.0384	0.1727	0.22	−0.3276	0.2348	−1.40
Female × category II	0.0235	0.1344	0.17	−0.0100	0.2002	−0.05
Female × category IIIB	−0.1425	0.3062	−0.47	0.3848	0.2498	1.54
Post-2009	−1.4076	0.1342	−10.49	−0.8825	0.1443	−6.11
Post-2009 × category IIIB	−0.6867	0.4691	−1.46	0.2033	0.4160	0.49
Constant	−2.6843	1.1101	−2.42	−4.7544	1.0302	−4.62
R^2	0.86			0.81		
Observations	136			136		
Pay ratio I	2.1778	0.3643	5.98	2.1836	0.4525	4.83
Pay ratio II	1.3584	0.3135	4.33	0.9617	0.4056	2.37
Pay ratio IIIA	0.9505	0.3269	2.91	0.4349	0.4274	1.02
Pay ratio IIIB	1.7938	0.6764	2.65	−0.2722	0.6733	−0.40

NOTE: Critical values for the *t*-statistic are 10 percent, 1.645; 5 percent, 1.96; and 1 percent, 2.58. SE = standard error.

Table B.9
Logit Regression of Share of Non–High School Diploma Graduate Accessions for Armed Forces Qualification Test Categories I Through IIIB, Marine Corps and Air Force

	Marine Corps			Air Force		
Variable	Coefficient	Standard Error	*t*	Coefficient	Standard Error	*t*
Pay ratio	0.9788	0.3574	2.74	0.8596	0.3670	2.34
Pay ratio × category I	1.5395	0.3718	4.14	–1.7537	0.4347	–4.03
Pay ratio × category II	0.1569	0.3017	0.52	–1.5182	0.3360	–4.52
Pay ratio × category IIIB	0.4459	0.4413	1.01	0.0545	0.4961	0.11
Recruiting goal	0.0486	0.0142	3.42	–0.0214	0.0100	–2.13
Deployment	–0.0587	0.0143	–4.12	–0.5471	0.0979	–5.59
Unemployment	–0.0625	0.0246	–2.54	0.2789	0.0342	8.17
Category I	–3.3109	0.6512	–5.08	3.3326	0.7776	4.29
Category II	–0.5595	0.5139	–1.09	3.1640	0.6116	5.17
Category IIIB	–1.0222	0.7194	–1.42	–0.0985	0.8062	–0.12
Female	–0.8535	0.1243	–6.86	–0.3317	0.1853	–1.79
Female × category I	–0.1652	0.2124	–0.78	0.6153	0.2371	2.60
Female × category II	0.0789	0.1619	0.49	0.3687	0.2030	1.82
Female × category IIIB	–0.1382	0.1934	–0.71	–0.1590	0.2586	–0.62
Post-2009	–0.9807	0.1753	–5.59	0.3187	0.1147	2.78
Post-2009 × category IIIB	–0.6629	0.2236	–2.97	–0.0551	0.2720	–0.20
Constant	–5.2164	0.7041	–7.41	–6.4074	0.7242	–8.85
R^2	0.76			0.73		
Observations	136			136		
Pay ratio I	2.5184	0.3868	6.51	–0.8940	0.4196	–2.13
Pay ratio II	1.1357	0.3099	3.66	–0.6585	0.2888	–2.28
Pay ratio IIIA	0.9788	0.3574	2.74	0.8596	0.3670	2.34
Pay ratio IIIB	1.4248	0.3760	3.79	0.9142	0.3836	2.38

NOTE: Critical values for the *t*-statistic are 10 percent, 1.645; 5 percent, 1.96; and 1 percent, 2.58. SE = standard error.

APPENDIX C

Why Not Compare Basic Pay to the Employment Cost Index?

Comparing basic pay to the ECI raises several problems. First, basic pay is not as broad a measure of pay as RMC, and RMC and basic pay do not necessarily change at the same rate. Basic pay and RMC did change at the same rate for many years (Figure C.1), and, for those years, a comparison of the ECI to basic pay was essentially a comparison of the ECI to RMC. Since FY 2000, however, RMC has increased faster than basic pay. This was driven by increases in the

Figure C.1
Indexes of Regular Military Compensation and Basic Pay, 1973 to 2016

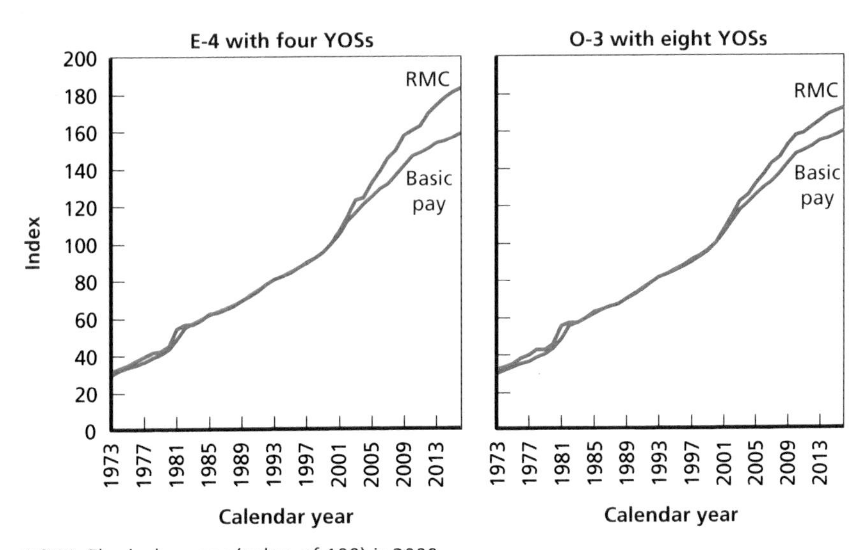

NOTE: The index year (value of 100) is 2000.

housing allowance, which expanded to cover the entire expected cost of housing deemed appropriate to a military family given its location, pay grade, and dependency status. ECI–basic pay comparisons therefore now exaggerate any gap and might give the appearance of a gap when none exists.

Second, the size of the ECI–basic pay gap depends on the base year. Tables C.1 and C.2 show the percentage increases in current-year dollars in basic pay, RMC, and the ECI using different base years for an E-4 with four YOSs and an O-3 with eight YOSs, respectively. The base years are 1982, the year when the second of two large increases in basic pay helped to restore it to parity with civilian pay; 1990, prior to the military drawdown; 2000, the year of a structural increase in basic pay and the beginning of a decade of higher-than-usual annual increases in basic pay and increases in BAH; and 2010, the final year of the higher-than-usual increases in basic pay and the start of a period with smaller-than-usual basic-pay increases in FY 2014, FY 2015, and FY 2016.[1]

When 1982 is the base year, by 1990, there is a 7-percentage-point pay gap for basic pay—the ECI has grown by 38 percent, versus 31 percent for basic pay. RMC did not grow as fast as basic pay in this period, so the ECI–RMC gap is wider in this case at 11 percentage points. Basic pay grew slower than the ECI during the 1990s, producing a gap of 10 percentage points by 2000 despite the higher-than-ECI increase in basic pay in FY 2000 (a 4.8-percent increase in basic pay versus a 4.3-percent increase in lagged ECI). The ECI–RMC gap in 2000 was 15 percentage points (92 – 77) for the E-4 and 17 percentage points (92 – 75) for the O-3 because of slow adjustments to BAH. The basic pay and RMC increases in the 2000s turned the situation around. By

---

[1] Figure C.1 and Tables C.1 and C.2 use the 12-month percentage change in the ECI for private industry wages and salaries as of September of the year shown. This provides a *contemporaneous* comparison of whether military pay is keeping up with civilian pay. We note, however, that the statutory guidance uses a lagged ECI for adjusting basic pay. It is the percentage change in ECI for the 12-month period ending the September that comes 15 months before the January 1 when the enacted basic pay takes effect. The lagged ECI is available for use at the start of budget preparations for the coming FY. We also used 1984 as the base year for basic pay and RMC and used 1982 for ECI, and the results were similar.

Table C.1
Percentage Increases in Basic Pay, Regular Military Compensation, and the Employment Cost Index for an E-4 with Four Years of Service, in Current-Year Dollars

	From 1982			From 1990			From 2000			From 2010		
To Year	Basic Pay	RMC	ECI	Basic Pay	RMC	ECI	Basic Pay	RMC	ECI	Basic Pay	RMC	ECI
1990	31	27	38									
2000	82	77	92	39	40	39						
2010	168	185	150	104	125	57	47	61	30			
2016	190	224	182	121	156	105	49	63	47	8	14	13

Table C.2
Percentage Increases in Basic Pay, Regular Military Compensation, and the Employment Cost Index for an O-3 with Eight Years of Service, in Current-Year Dollars

	From 1982			From 1990			From 2000			From 2010		
To Year	Basic Pay	RMC	ECI	Basic Pay	RMC	ECI	Basic Pay	RMC	ECI	Basic Pay	RMC	ECI
1990	31	27	38									
2000	82	75	92	39	38	39						
2010	168	175	150	104	117	82	47	57	30			
2016	190	201	182	121	138	105	59	72	47	8	9	13

2010, the basic-pay increase was 18 percentage points greater than the ECI's increase for both the E-4 and the O-3, and the RMC increase was 35 percentage points greater for the E-4 and 25 percentage points greater for the O-3. The slowed rate of increase in basic pay since 2010 narrowed the basic-pay edge to an 8-percentage-point difference as of 2016, but increases in BAH led RMC to be 42 percentage points (= 224 – 182) ahead of the ECI for the E-4 and 19 percentage points (201 – 182) ahead of the ECI for the O-3.[2]

When 2010 is the base year, Tables C.1 and C.2 show that, for both enlisted personnel and officers, basic pay had grown by 8 percent by 2016, which compares to a 13-percent increase in the ECI. This implies a pay gap of 5 percentage points. However, our analysis of the RMC percentile by YOS found virtually no change in either the enlisted or officer percentile between 2009 and 2016.

Changes from other base years produced different results. This is not surprising but is useful to keep in mind if a factor in selecting a base year is how well recruiting and retention were going. For instance, there were recruiting and retention problems around 2000 during the dot.com boom; yet, if 1990 were the base year, there would be no apparent pay gap in 2000.

Third, pay-gap comparisons are in current-year dollars. Tables C.1 and C.2 are in current-year dollars to relate to typical pay-gap esti-

---

[2] Here is an example from 2016 of an ECI–basic pay comparison that misses the contribution of BAH, which is present in RMC:

> The 2016 raise has something else in common with the minimal 2014 and 2015 increases: It does not keep pace with average private-sector wage growth last year, which was 1.8 percent—making 2014, 2015 and 2016 the only three years since 1999 that the military basic pay raise lagged average civilian wage growth. That is reviving concerns over the so-called "pay gap" between military and private-sector wages. Under that controversial concept—which military officials have long argued is not a truly representative measure—the gap peaked around 1998–99 at 13.5 percent, a time when military recruiting and retention rates were suffering considerably across the services. Congress and the Pentagon agreed to a series of higher-than-average military raises over the first few years of the new millennium, narrowing the gap to a low of 2.4 percent. But relatively stable recruiting and retention in recent years have led lawmakers and defense officials to ease off on big military raises, with the result that the gap between average military and private-sector pay, according to those who track it, has grown once again to about 5 percent this year. ("2016," 2016)

mates, but current-year dollars exaggerate the optics of the pay gap. Tables C.3 and C.4 adjust for inflation and show, as expected, that the gap or gain is smaller when viewed in real dollars.[3] For instance, the 10-percentage-point ECI–basic pay gap in 2000 using 1982 as the base year becomes a 5-percentage-point gap after the inflation adjustment. RMC, instead of being 42 percentage points ahead of the ECI in 2016 for the E-4, becomes 18 percentage points ahead.

A fourth problem with ECI-based comparisons is that the ECI does not capture differences in the rate of civilian pay growth between age/education groups. This is important because the age/education mix differs substantially between military personnel and civilian workers and the annual rate of change in civilian pay for age/education groups differs by education. Take 28- to 32-year-old men, for example. We estimated real wage *decreases* from 2000 through 2015 of 0.95 percent per year (i.e., slightly less than 1 percent) for high school graduates, 0.76 percent per year for those with some college, 0.77 for bachelor's, and 0.45 percent per year for those with master's degrees or higher (Appendix A). The some-college and bachelor's estimates of the wage decrease are statistically identical, but other pairwise comparisons are significantly different. The ECI, which is not designed to capture differential rates of wage change by age/education, misses these differences across groups.

[3] The percentage increases in Tables C.3 and C.4 are based on inflation-adjusted values of basic pay, RMC, and ECI using the Consumer Price Index for All Urban Consumers. The ECI is itself an index, so it might seem strange to adjust it for inflation. But the ECI is an index of the market-basket cost of labor, so adjusting it by the Consumer Price Index for All Urban Consumers yields the cost of labor relative to the purchasing power of the dollar with respect to a market basket of goods.

Table C.3
Percentage Increases in Basic Pay, Regular Military Compensation, and the Employment Cost Index for an E-4 with Four Years of Service, in 2014 Dollars

	From 1982			From 1990			From 2000			From 2010		
To Year	Basic Pay	RMC	ECI	Basic Pay	RMC	ECI	Basic Pay	RMC	ECI	Basic Pay	RMC	ECI
1990	–3	–7	2									
2000	2	–1	7	5	6	6						
2010	18	26	11	22	35	9	16	27	3			
2016	20	34	16	24	43	15	14	25	8	1	6	5

Table C.4
Percentage Increases in Basic Pay, Regular Military Compensation, and the Employment Cost Index for an O-3 with Eight Years of Service, in 2014 Dollars

	From 1982			From 1990			From 2000			From 2010		
To Year	Basic Pay	RMC	ECI	Basic Pay	RMC	ECI	Basic Pay	RMC	ECI	Basic Pay	RMC	ECI
1990	–3	–7	2									
2000	2	–2	7	5	5	6						
2010	19	22	11	22	30	9	16	24	3			
2016	20	25	16	24	33	15	17	27	8	1	2	5

APPENDIX D

A Two-Goal Model of Recruitment Resource Allocation

Motivation for the Model

Empirical analyses of recruiting generally focus on the number of high-quality recruits.[1] One rationale for this focus is that high-quality recruits are not demand constrained; the service will take as many as it can recruit. Therefore, the effects that recruiting resources and military and civilian pay have on high-quality recruits can be interpreted as identifying the supply curve of these recruits. However, in the context of this research, we were interested in the overall cost of recruiting, which includes the cost of recruiting both high-quality and non–high-quality recruits. This broader scope is relevant because we wanted to consider the extent to which higher military pay than civilian pay affects the service's choice of recruiting resources and therefore the recruiting resource cost that the service incurs. The model described in this appendix will help to illustrate optimal recruiting input choices subject to the service's total contract mission, its goal for high-quality recruits within that mission, and factors outside the control of the recruiting command but nevertheless affecting the supply of both high-quality and non–high-quality recruits. These factors are military pay relative to civilian pay, represented by RMC, unemployment, and hostile deployment; RMC is of particular interest given the topic of this report. Subject to total recruiting mission, high-quality mission, RMC, unemployment, and deployment, the recruiting command's objective is to select the optimal level and mix of inputs under

[1] See Asch, Hosek, and Warner, 2007, for a survey of recruiting studies.

its control. These are recruiter time, bonuses, advertising, and waiver policy.[2]

To capture the overall recruiting budget, total mission and high-quality mission must both be considered. This requires a model that incorporates both of these objectives, and the mathematical technique for this is Kuhn–Tucker (K-T). In principle, recruiting inputs could be selected to exactly meet both objectives. But more generally, it could be that allocating enough resources to meet one of the objectives provides more than enough to meet the other objective. In this case, the other objective is "slack." K-T handles such situations. In the analysis described in this appendix, after setting up the problem, we considered two cases. In the first case, the two objectives are met; in the second case, one objective is met and the other objective is slack. The particular condition we examined is when the total mission is met and the high-quality mission is exceeded.

The model was intended to provide structured thought about how a service could respond to external recruiting conditions, and we tied the model back to RMC by reflecting on the following. In the first case, in which the total mission and high-quality mission are met exactly, we thought about how a service might respond when RMC is increased. Under the model's assumptions, an increase in RMC will increase the supply of all recruits, both high quality and non–high quality. Therefore, a service could decrease its recruiting inputs and still meet its total mission and its high-quality mission. In the second case, addressed below, the total mission is met exactly, and the high-quality mission is exceeded. This could correspond to circumstances in which RMC had increased but a service held its recruiting inputs constant. This case is consistent with the increase in the percentage of high-quality accessions seen in the Navy, Air Force, and Marine Corps in our study period. By comparison, as RMC increased, the Army did not increase its recruit quality, and the empirical evidence indicates some decrease in quality. This is consistent with decreasing recruiting inputs below what they otherwise would have been, as RMC increased.

[2] *Waiver policy* more broadly stands for eligibility policy and includes policy affecting the recruitment of prior-service personnel.

We also suggest another possible factor in Chapter Four, which is that the Army's recruit supply curve might have shifted back at the same rate or even faster than RMC increased. If so, recruit quality could not have been maintained or increased with an increase in recruiting inputs.

Recruiting Objectives

We assumed that the recruiting objectives were to achieve the total contract mission, m, with a certain percentage k of recruits being required to reach a quality goal set by DoD or the recruiting service. Defense guidance is that at least 60 percent of accessions be in categories I through IIIA and that at least 90 percent of accessions be tier 1 (Sellman, 2004). Our focus was on contracts. Although there is a difference between the number of enlistment contracts and the number of accessions, we did not extend the model to include DEP attrition. Some who sign contracts leave the DEP and do not enter a military service, so contracts have to be high enough to meet the accession goal.[3]

Production Functions

Consider two recruit contract production functions: one for high-quality recruits and the other for non–high-quality recruits. One could be for categories I–IIIA and the other for other categories, or one could represent HSDGs and the other non-HSDGs, or one could be for I–IIIA HSDGs and the other for non–I–IIIA HSDGs. We refer to the outputs as h and n. Achieving total mission m requires $h + n = m$, and a goal of k-percent high quality implies $h \geq km$. The services seek to

[3] Data for the Army indicate a lower rate of attrition from DEP for category I–IIIA high school graduates than for others (Knapp, Orvis, et al., 2018), so the 60-percent guidance for contracts might be conservative. For example, if 58 percent of contracts were in categories I through IIIA, category I–IIIA accessions might be 62 percent. In any case, the contract mission can be adjusted to allow for DEP attrition.

achieve total mission but, to keep the inflow of personnel in balance over time, do not want to exceed it. So, $n \leq (1 - k)m$.

Each function has specific inputs and common inputs. The specific inputs are recruiter time and average enlistment bonus (in this appendix, just "bonus" for short), which is the product of the probability of receiving a bonus and the expected amount given receipt. Recruiter time and bonus vary separately in each production function. The common inputs are advertising, waiver policy, relative military pay (RMC percentile), and unemployment rate. The level of a common input is the same in both production functions. However, a common input's effect on output can differ between the production functions, as can the effects of recruiter time and bonus.

The recruiting command bears the cost of recruiters, bonuses, and advertising. It does not bear the cost of RMC apart from recruiter compensation, and there is no monetary cost of waivers or the unemployment rate.

The production inputs are recruiter time (r), bonus (b), advertising (a), waiver guidance (w), and input x, which can represent RMC, unemployment, or deployment. The recruiting command can vary r, b, a, and, to a limited extent, w but cannot affect x. Let c be total cost borne by the recruiting command, $p1$ the price of a unit of recruiter time, $p2$ the price of a bonus, and $p3$ the price of a unit of advertising (e.g., 1,000 impressions). The production functions and cost are

$$h = h[r1, b1, a, w, x]$$
$$n = n[r2, b2, a, w, x]$$
$$c = p1(r1 + r2) + p2(b1h + b2n) + p3a.$$

A change in w or x acts to shift the supply of potential recruits and has the effect of increasing the productivity of the inputs that the recruiting command can vary. Waivers can be varied at no cost but only over a small range. An increase in waivers increases the potential supply of recruits by increasing the eligibility of youths who would otherwise be ineligible. An increase in RMC or unemployment shifts out the entire supply curve. Another variable might affect the recruiting climate: the extent of personnel deployment to hostile areas. Deployment's impact on the recruiting climate could be positive or negative, depending on the perceived merit of the cause and the danger and sacrifice likely to be faced.

The Problem and First-Order Conditions

The recruiting command wants to recruit at least *km* of *h*, $0 < k < 1$, and a total of $h + n = m$ recruits, at least cost. The problem can be set up as a K-T problem with an inequality constraint for *h* and an equality constraint for meeting the overall goal, *m*. The problem also includes a constraint for the limited flexibility to vary waivers. The classic K-T setup is to maximize a function subject to constraints expressed as $g(x) \leq c$, where (here) *x* is a vector of variables. We wanted to minimize cost, which is the same as maximizing the negative of cost, subject to constraints in the form $g(x) \geq c$ (e.g., *h* must be greater than or equal to *km*):

$$\begin{aligned} L[r1,r2,b1,b2,a,w,x] = -\big(p1(r1+r2) + p2(b1h+b2n) + p3a\big) \\ +\lambda 1\big(h[r1,b1,a,w,x]-km\big) \\ +\lambda 2\big(n[r2,b2,a,w,x]-(m-h[r1,b1,a,w,x])\big) \\ +\lambda 3(w - w\max), \end{aligned}$$

where $\lambda 1 \geq 0$, $\lambda 2 > 0$, and $\lambda 3 \geq 0$. The strict inequality for $\lambda 2$ will require that, at the optimum, the total mission *m* is met exactly.

The first-order conditions for cost minimization are

$$
\begin{aligned}
&L_{r1} = -p1 + (\lambda 1 + \lambda 2) h_{r1} = 0 \\
&L_{b1} = -p2(h + h_{b1}) + (\lambda 1 + \lambda 2) h_{b1} = 0 \\
&L_{r2} = -p1 + \lambda 2 n_{r2} = 0 \\
&L_{b2} = -p2(n + n_{b2}) + \lambda 2 n_{b2} = 0 \\
&L_{a} = -p3 + (\lambda 1 + \lambda 2) h_{a} + \lambda 2 n_{a} = 0 \\
&L_{w} = (\lambda 1 + \lambda 2) h_{w} + \lambda 2 n_{w} = 0 \\
&\lambda 1(km - h[r1, b1, a, w, x]) = 0 \\
&\lambda 2(m - h[r1, b1, a, w, x] - n[r2, b2, a, w, x]) = 0 \\
&\lambda 3(w - w\max) = 0.
\end{aligned}
$$

There are two solutions: (1) $h = km$ and $n = m - h = (1 - k)m$, with $h + n = m$; and (2) $h > km$, and $n < (1 - k)m$, with $h + n = m$. Under (1), the second- and third-to-last conditions are satisfied by $\lambda 1 > 0$, $\lambda 2 > 0$, $h[r1,b1,a,w,x] = km$, and $n\,[r2,b2,a,w,x] = (1 - k)m$. Under (2), satisfying the third-to-last condition requires $\lambda 1 = 0$ given that $h > km$. Also, because $\lambda 2 > 0$ in the problem setup, the second-to-last condition could alternatively be expressed as the expression in parentheses equaling 0.

Case 1: Both Goals Are Met Exactly

In this case, both $\lambda 1$ and $\lambda 2$ are positive and the production constraints are satisfied with equality.

Recruiter Time

The condition for recruiter time in the production of category I–IIIAs, $r1$, is $p1 = (\lambda 1 + \lambda 2)h_{r1}$. Here, an increase in h contributes toward reaching both constraints, and $(\lambda 1 + \lambda 2)h_{r1}$ can be interpreted as marginal value of this dual contribution. $\lambda 1$ is the marginal cost of h and $\lambda 2$ is the marginal cost of m when first- and second-order conditions are met. Marginal cost will be constant with respect to output if there are constant returns to scale and input prices are given, for example, or increasing if returns are less than to scale, as might be expected.

Offsetting the possible increase in marginal cost is the diminishing marginal product of the input. At an optimum, although the marginal cost of output might be increasing in *h* and *m*, recruiter time will have increased to the point at which its marginal product has diminished enough for the combined term, $(\lambda 1 + \lambda 2)h_{r1}$, to just equal the price of recruiter time, $p1$.

The condition for recruiter time in the production of *n* is that price equals $\lambda 2$ (the marginal cost of *n* at the optimum) times the marginal product of $r2$ in producing *n*.

Looking at the recruiter time conditions together, because both conditions equal 0, we have

$$-p1+(\lambda 1+\lambda 2)h_{r1} = -p1+\lambda 2 n_{r2}$$
$$(\lambda 1+\lambda 2)h_{r1} = \lambda 2 n_{r2}$$
$$\lambda 1 h_{r1} = \lambda 2(n_{r2} - h_{r1}).$$

This condition implies that the marginal product of recruiter time in *h* is less than in *n*. This follows because $\lambda 1 > 0$, $h_{r1} > 0$, so the left-hand side is positive. Also, $\lambda 2 > 0$, so for the right-hand side to be positive, it must be that $n_{r2} - h_{r1} > 0$. This is consistent with the idea that recruiter time would be greater in the production of *h* because an additional *h* contributes to meeting both goals; with decreasing returns, more recruiter time in the production of *h* would cause the marginal product of recruiter time to be lower in the production of *h* than in the production of *n*.

Bonus

In the first-order condition for $b1$, the term $p2(h + h_{b1})$ is the marginal cost of the bonus. It equals the price of a bonus times the number initially receiving it plus an adjustment, h_{b1}, for the incremental increase in that number when the bonus is raised by one unit. This is equal to the marginal value product of the bonus, $(\lambda 1 + \lambda 2)h_{b1}$. Because the bonus is a dollar amount, it would seem natural to designate the unit price of a bonus as \$1. But the cost of committing to pay \$1 of enlistment bonus is less than \$1 because the bonus is not paid immediately, and part of the bonus might be paid on installment, decreasing the present cost of the bonus. Also, not all bonus recipients stay in service long enough to collect all installments. Another type of bonus, a quick-ship bonus, is paid to enlistees willing to enter basic training within 60 days of signing the enlistment contract.[4] Treating enlistment and quick-ship bonuses together, we expected $p2$ to be less than 1.

The bonus conditions for h and n production imply that

$$\frac{p2(h+h_{b1})}{p2(n+n_{b2})}=\frac{(\lambda 1+\lambda 2)h_{b1}}{\lambda 2 n_{b2}}.$$

Cancelling $p2$ and dividing through by the marginal products gives

$$\frac{\frac{h}{h_{b1}}+1}{\frac{n}{n_{b2}}+1}=\frac{\lambda 1}{\lambda 2}+1.$$

The right-hand side is positive. For a higher ratio of $\lambda 1$ to $\lambda 2$, the left-hand side must also have a higher ratio. This can be achieved by, for instance, increasing $b1$ such that its marginal product h_{b1} decreases by a sufficient amount to increase the ratio. More generally, the higher $\lambda 1$ to $\lambda 2$, the greater the h bonus relative to the n bonus. This is consistent with greater difficulty (higher marginal cost) in recruiting h leading to a higher h bonus (i.e., a higher bonus amount, a higher fraction of recruits receiving a bonus, or both).

Taking the first-order conditions for recruiter time and bonus for h together,

$$\frac{p1}{p2(n+n_{b2})}=\frac{h_{r1}}{h_{b1}}.$$

---

[4] In the Army, for instance, all bonuses are paid out in anniversary payments unless the bonus is below $10,000, in which case it is paid at the completion of initial entry training.

This is the well-known optimality condition that the ratio of input marginal costs equals the ratio of their marginal products. A similar result applies for recruiter time $r2$ and bonus $b2$ in producing n.

Advertising

Purchasing one unit of advertising increases impressions among both h and n youth populations and their influencers, which affects h and n recruiting. The first-order condition sets price equal to the marginal cost of h and n times the respective marginal product of advertising in producing h and n:

$$p3 = (\lambda 1 + \lambda 2)h_a + \lambda 2\, n_a.$$

How does this condition relate to recruiter time? If we consider adding one unit of recruiter time to both h and n recruiting, adding their first-order conditions gives $2\,p1 = (\lambda 1 + \lambda 2)h_{r1} + \lambda 2\, n_a$. Thus,

$$\frac{p3}{2\,p1} = \frac{(\lambda 1+\lambda 2)h_a + \lambda 2 n_a}{(\lambda 1+\lambda 2)h_{r1} + \lambda 2 n_{r2}}.$$

At the optimum, the ratio of the price of advertising to the price of two units of recruiter time equals the ratio of the marginal cost–weighted marginal products. Intuitively, the ratio is based on two units of recruiter time because, given that advertising is a common input affecting both h and n, an apt comparison involves changing recruiter time in both h and n production.

Waivers

Because waivers are costless to the recruiting command, the first-order condition, $(\lambda 1 + \lambda 2)h_w + \lambda 2\, n_w = 0$, indicates that waivers would be increased until their marginal product was 0. This is misleading, however, because their range is limited. Waivers can be increased only to the maximum allowable level, $w$max, as reflected in the last first-order condition. When that constraint is binding, the marginal products of waivers would be positive, resulting in a corner solution and requiring $\lambda 3 = 0$ to satisfy the first-order condition.

Regular Military Compensation and Unemployment

These variables affect the supply of recruits and are beyond control of the recruiting command. They do not affect the price of recruiter time, bonus, or advertising but can be expected to affect their marginal products. An increase in RMC or unemployment shifts out the supply curve of recruits or, stated differently, increases the probability that an individual is willing to join the military. Therefore, at any given level of r, b, a, and w, an increase in RMC or unemployment should increase their marginal product and decrease the marginal cost of h and n recruits. This would perturb the first-order conditions. Consider the effect on the condition for $r1$, which is $-p1 + (\lambda 1 + \lambda 2)h_{r1} = 0$. Now increase x, which represents RMC or unemployment:

$$\frac{\partial\left(-p1+(\lambda 1+\lambda 2)h_{r1}\right)}{\partial x}=\frac{\partial(\lambda 1+\lambda 2)}{\partial x}h_{r1}+(\lambda 1+\lambda 2)\frac{\partial h_{r1}}{\partial x}=0$$

$$\frac{\frac{\partial(\lambda 1+\lambda 2)}{\partial x}}{\frac{\partial h_{r1}}{\partial x}}=\frac{d(\lambda 1+\lambda 2)}{dh_{r1}}=-\frac{\lambda 1+\lambda 2}{h_{r1}}.$$

The first-order condition equals 0, and the first line in the equation says that the change in the first-order condition with respect to a change in x must equal 0 to preserve optimality after the change. The second line applies the implicit function theorem. The result indicates the trade-off between the marginal cost of h and the marginal product of $r1$. Because the right-hand side of the second line is negative, the increase in the marginal product of $r1$ from the increase in x must be accompanied by a decrease in $\lambda 1 + \lambda 2$ to maintain equilibrium. Also, the equilibrium quantity of recruiter time will decrease because it is more productive, yet its price has not changed; less recruiter time is needed to produce the same output. For similar reasons, the equilibrium quantities of $b1$ and advertising will decrease.

Case 2: *H* Mission Is Exceeded, and *M* Mission Is Met Exactly

In this case, the K-T conditions require that the Lagrange multiplier on the *h* constraint is 0 ($\lambda 1 = 0$) because the condition is slack. In effect, adding further *h* beyond the goal level *km* adds no value toward meeting the *h* constraint, but it does contribute toward the *m* constraint. When $\lambda 1 = 0$, the first-order conditions are

$$L_{r1} = -p1 + \lambda 2 h_{r1} = 0$$
$$L_{b1} = -p2(h + h_{b1}) + \lambda 2 h_{b1} = 0$$
$$L_{r2} = -p1 + \lambda 2 n_{r2} = 0$$
$$L_{b2} = -p2(n + n_{b2}) + \lambda 2 n_{b2} = 0$$
$$L_{a} = -p3 + \lambda 2 h_{a} + \lambda 2 n_{a} = 0$$
$$L_{w} = \lambda 2 h_{w} + \lambda 2 n_{w} = 0$$
$$\lambda 2(m - h[r1, b1, a, w, x] - n[r2, b2, a, w, x]) = 0$$
$$\lambda 3(w - w\max) = 0.$$

The discussion of case 1 applies, but now $\lambda 1$ no longer figures in the conditions. Because $0 < \lambda 2 = \lambda 1 + \lambda 2$, attaining the first-order conditions requires that the marginal products of *r*1, *b*1, *a*, and *w* be higher than in case 1. For example, the first-order condition for *r*1 is $p1 = (\lambda 1 + \lambda 2)h_{r1}$ in case 1 and $p1 = \lambda 2\ h_{r1}$ in case 2, so, with price *p*1 unchanged, the marginal product h_{r1} must be higher to satisfy the condition. Given that inputs have diminishing marginal products, obtaining a higher marginal product requires less input usage than in case 1.

When is case 2 relevant? The broad answer is when recruiting conditions are favorable and a service has little difficulty reaching the goal for *h*. This is when (1) RMC percentile or unemployment is high (or both are high); (2) waiver guidance provides relatively liberal eligibility to qualify for military service; or (3) recruiting resources are maintained at high levels. The latter might be due to inefficiency, budgeting frictions preventing resource reallocation, or uncertainty regarding future recruiting environments.[5]

Two other points are important. The services differ in their appeal to prospective recruits, yet the RMC percentile and unemployment are the same across services. For given RMC and unemployment, some services could be in case 1 while others are in case 2. Moreover, this depends critically on mission levels. DoD guidance is at least 60 percent categories I–IIIA and at least 90 percent tier 1, but a service can set its own goals as long as they are not below DoD guidance. A service might want 70 percent categories I–IIIA or 95 percent HSDGs. It might be less costly for an "attractive" service to meet a 70-percent category I–IIIA goal, for instance, than for a less attractive service to meet a 60-percent category I–IIIA goal—the recruiting cost function differs by service.

The model provides some insight into why the Army might not have increased recruit quality as RMC increased between 1999 and 2015. The model also shows the optimal resourcing conditions subject to meeting the quantity goal and exceeding the quality goal. As we discuss in Chapter Four, it is possible that the Army set quality goals and programmed recruiting resources to sustain, but not increase, recruit quality because the incremental benefit relative to the incremental cost of HSDG II–IIIB recruits was lower than that of category I–IIIB non-HSDG recruits. The model shows that an increase in RMC could lead to a decrease in recruiting resources, such as recruiter effort and bonuses, subject to meeting the quality and quantity goals. The model also shows that, had they maintained their levels of recruiting inputs, the Navy, Air Force, and Marine Corps could have increased recruit quality as RMC increased.

---

[5] For example, in 2010 and 2011, the recruiting environment was excellent, but recruiter levels for the Army remained high.

References

"2001 US Military Basic Pay Charts," Navy CyberSpace, undated. As of July 31, 2018:
https://www.navycs.com/charts/2001-military-pay-chart.html

"2016: Another Lukewarm Year for Military Compensation," *Military Times*, January 18, 2016. As of July 31, 2018:
https://www.militarytimes.com/pay-benefits/military-pay-center/2016/01/18/2016-another-lukewarm-year-for-military-compensation/

Accession Policy Directorate, Office of the Under Secretary of Defense for Personnel and Readiness, "Accession Goals, Achievement and Quality," 2017, not available to the general public.

Air Force Institute of Technology, "About the Graduate School," last updated July 27, 2018. As of July 31, 2018:
https://www.afit.edu/EN/page.cfm?page=134

Asch, Beth J., Paul Heaton, James Hosek, Paco Martorell, Curtis Simon, and John T. Warner, *Cash Incentives and Military Enlistment, Attrition, and Reenlistment*, Santa Monica, Calif.: RAND Corporation, MG950-OSD, 2010. As of May 19, 2017:
https://www.rand.org/pubs/monographs/MG950.html

Asch, Beth J., James Hosek, and John T. Warner, *An Analysis of Pay for Enlisted Personnel*, Santa Monica, Calif.: RAND Corporation, DB344-OSD, 2001. As of March 6, 2018:
https://www.rand.org/pubs/documented_briefings/DB344.html

———, "New Economics of Manpower in the Post–Cold War Era," in Todd Sandler and Keith Hartley, eds., *Handbook of Defense Economics*, Vol. 2, Elsevier, 2007, pp. 1075–1138.

Asch, Beth J., John A. Romley, and Mark E. Totten, *The Quality of Personnel in the Enlisted Ranks*, Santa Monica, Calif.: RAND Corporation, MG324-OSD, 2005. As of November 23, 2017:
https://www.rand.org/pubs/monographs/MG324.html

Bedno, Sheryl A., Christine E. Lang, William E. Daniell, Andrew R. Wiesen, Bennett Datu, and David W. Niebuhr, "Association of Weight at Enlistment with Enrollment in the Army Weight Control Program and Subsequent Attrition in the Assessment of Recruit Motivation and Strength Study," *Military Medicine*, Vol. 175, No. 3, March 2010, pp. 188–193. As of August 28, 2018: http://www.amsara.amedd.army.mil/Documents/ARMS_Publication/2.%20Bedno-Mar-2010.pdf

Buddin, Richard, *Analysis of Early Military Attrition Behavior*, Santa Monica, Calif.: RAND Corporation, R3069-MIL, 1984. As of November 10, 2017: https://www.rand.org/pubs/reports/R3069.html

———, *Success of First-Term Soldiers: The Effects of Recruiting Practices and Recruit Characteristics*, Santa Monica, Calif.: RAND Corporation, MG262-A, 2005. As of December 21, 2017:
https://www.rand.org/pubs/monographs/MG262.html

Bureau of Labor Statistics, U.S. Department of Labor, "Databases, Tables and Calculators by Subject," undated. As of August 30, 2018:
https://www.bls.gov/data/

Chapman, Chris, Jennifer Laird, Nicole Ifill, and Angelina KewalRamani, *Trends in High School Dropout and Completion Rates in the United States: 1972–2009—Compendium Report*, Washington, D.C.: U.S. Department of Education, National Center for Education Statistics, Institute of Education Sciences, NCES 2012-006, October 2011. As of January 13, 2018:
https://nces.ed.gov/pubs2012/2012006.pdf

Christensen, Garret, "Occupational Fatalities and the Labor Supply: Evidence from the Wars in Iraq and Afghanistan," *Journal of Economic Behavior and Organization*, Vol. 139, July 2017, pp. 182–195.

CNA Analysis and Solutions, "Population Representation in the Military Services," undated. As of May 19, 2017:
https://www.cna.org/research/pop-rep

Defense Travel Management Office, "Basic Allowance for Housing (BAH) Frequently Asked Questions," updated January 27, 2015. As of July 31, 2018: http://www.defensetravel.dod.mil/site/faqbah.cfm

———, *A Primer on the Basic Allowance for Housing (BAH) for the Uniformed Services*, January 2018. As of July 31, 2018:
http://www.defensetravel.dod.mil/Docs/perdiem/BAH-Primer.pdf

Dertouzos, James N., *Recruiter Incentives and Enlistment Supply*, Santa Monica, Calif.: RAND Corporation, R3065-MIL, 1985. As of May 19, 2017:
http://www.rand.org/pubs/reports/R3065.html

Directorate of Compensation, Office of the Under Secretary of Defense for Personnel and Readiness, *Selected Military Compensation Tables*, Washington, D.C., 1999.

———, *Selected Military Compensation Tables*, Washington, D.C., c. 2000.

———, *Selected Military Compensation Tables*, Washington, D.C., c. 2001.

———, *Selected Military Compensation Tables*, Washington, D.C., c. 2002.

———, *Selected Military Compensation Tables*, Washington, D.C., c. 2003. As of August 5, 2018:
https://militarypay.defense.gov/Portals/3/Documents/Reports/greenbook_fy2003.pdf

———, *Selected Military Compensation Tables*, Washington, D.C., January 1, 2004. As of August 5, 2018:
https://militarypay.defense.gov/Portals/3/Documents/Reports/greenbook_fy04.pdf

———, *Selected Military Compensation Tables*, Washington, D.C., January 1, 2005. As of August 5, 2018:
https://militarypay.defense.gov/Portals/3/Documents/Reports/greenbook_fy05.pdf

———, *Selected Military Compensation Tables*, Washington, D.C., January 1, 2006. As of August 5, 2018:
https://militarypay.defense.gov/Portals/3/Documents/Reports/greenbook2_fy06.pdf

———, *Selected Military Compensation Tables*, Washington, D.C., April 1, 2007. As of August 5, 2018:
https://militarypay.defense.gov/Portals/3/Documents/Reports/GreenBook_APRIL_40YOS_2007_Dist.pdf

———, *Selected Military Compensation Tables*, Washington, D.C., January 1, 2008. As of August 5, 2018:
https://militarypay.defense.gov/Portals/3/Documents/Reports/GreenBook_2008.pdf

———, *Selected Military Compensation Tables*, Washington, D.C., January 1, 2009. As of August 5, 2018:
https://militarypay.defense.gov/Portals/3/Documents/Reports/GreenBook_2009.pdf

———, *Selected Military Compensation Tables*, Washington, D.C., January 1, 2010. As of August 5, 2018:
https://militarypay.defense.gov/Portals/3/Documents/Reports/GreenBook_2010.pdf

———, *Selected Military Compensation Tables*, Washington, D.C., January 1, 2011. As of August 5, 2018:
https://militarypay.defense.gov/Portals/3/Documents/Reports/GreenBook_2011.pdf

———, *Selected Military Compensation Tables*, Washington, D.C., January 1, 2012. As of August 5, 2018:
https://militarypay.defense.gov/Portals/3/Documents/Reports/GreenBook_2012.pdf

———, *Selected Military Compensation Tables*, Washington, D.C., January 1, 2013. As of August 5, 2018:
https://militarypay.defense.gov/Portals/3/Documents/Reports/GreenBook_2013.pdf

———, *Selected Military Compensation Tables*, Washington, D.C., January 1, 2014. As of August 5, 2018:
https://militarypay.defense.gov/Portals/3/Documents/Reports/GreenBook_2014.pdf

———, *Selected Military Compensation Tables*, Washington, D.C., January 1, 2015. As of August 5, 2018:
https://militarypay.defense.gov/Portals/3/Documents/Reports/GreenBook_2015.pdf

———, *Selected Military Compensation Tables*, Washington, D.C., January 1, 2016. As of August 4, 2018:
https://militarypay.defense.gov/Portals/3/Documents/Reports/GreenBook_2016.pdf

———, *Selected Military Compensation Tables*, Washington, D.C., January 1, 2017. As of December 30, 2017:
http://militarypay.defense.gov/Portals/3/Documents/Reports/GreenBook_2017.pdf?ver=2017-04-24-232444-753

DoD—*See* U.S. Department of Defense.

Drasgow, Fritz, University of Illinois at Urbana-Champaign, *Tailored Adaptive Personality Assessment System (TAPAS)*, briefing presented to the International Personnel Assessment Council, July 22, 2013. As of August 28, 2018:
http://www.ipacweb.org/Resources/Documents/conf13/drasgow.pdf

Eighmey, John, "Why Do Youth Enlist? Identification of Underlying Themes," *Armed Forces and Society*, Vol. 32, No. 2, 2006, pp. 307–328.

Federal Reserve Bank of St. Louis, "Employment Cost Index: Wages and Salaries—Private Industry Workers," undated (a). As of July 31, 2018:
https://fred.stlouisfed.org/series/ECIWAG

———, "Real Median Household Income in the United States," undated (b). As of July 31, 2018:
https://fred.stlouisfed.org/series/MEHOINUSA672N

Granger, C. W. J., and P. Newbold, "Spurious Regressions in Economics," *Journal of Econometrics*, Vol. 2, No. 2, July 1974, pp. 111–120.

Heffner, Tonia S., Roy C. Campbell, and Fritz Drasgow, *Select for Success: A Toolset for Enhancing Soldier Accessioning*, U.S. Army Research Institute for the Behavioral and Social Sciences, Special Report 70, March 2011. As of August 28, 2018: http://www.dtic.mil/dtic/tr/fulltext/u2/a554057.pdf

Hosek, James, and Paco Martorell, *How Have Deployments During the War on Terrorism Affected Reenlistment?* Santa Monica, Calif.: RAND Corporation, MG873-OSD, 2009. As of November 23, 2017: https://www.rand.org/pubs/monographs/MG873.html

Hosek, James, Christine E. Peterson, Jeannette Van Winkle, and Hui Wang, *A Civilian Wage Index for Defense Manpower*, Santa Monica, Calif.: RAND Corporation, R4190-FMP, 1992. As of March 6, 2018: https://www.rand.org/pubs/reports/R4190.html

Kapp, Lawrence, *Recruiting and Retention: An Overview of FY2011 and FY2012 Results for Active and Reserve Component Enlisted Personnel*, Congressional Research Service, RL32965, May 10, 2013. As of March 11, 2018: https://fas.org/sgp/crs/natsec/RL32965.pdf

Kilburn, M. Rebecca, and Jacob Alex Klerman, *Enlistment Decisions in the 1990s: Evidence from Individual-Level Data*, Santa Monica, Calif.: RAND Corporation, MR944-OSD/A, 1999. As of July 27, 2018: https://www.rand.org/pubs/monograph_reports/MR944.html

Knapp, David, Beth J. Asch, Michael G. Mattock, and James Hosek, *An Enhanced Capability to Model How Compensation Policy Affects U.S. Department of Defense Civil Service Retention and Cost*, Santa Monica, Calif.: RAND Corporation, RR1503-OSD, 2016. As of July 27, 2018: https://www.rand.org/pubs/research_reports/RR1503.html

Knapp, David, Bruce R. Orvis, Christopher E. Maerzluft, and Tiffany Tsai, *Resources Required to Meet the U.S. Army's Enlisted Recruiting Requirements Under Alternative Recruiting Goals, Conditions, and Eligibility Policies*, Santa Monica, Calif.: RAND Corporation, RR2364-A, 2018. As of July 27, 2018: https://www.rand.org/pubs/research_reports/RR2364.html

Laurence, Janice H., *Education Standards for Military Enlistment and the Search for Successful Recruits*, Alexandria, Va.: Human Resources Research Organization, FR-PRD-84-4, February 1984. As of November 10, 2017: http://www.dtic.mil/dtic/tr/fulltext/u2/a139718.pdf

Loughran, David S., and Bruce R. Orvis, *The Effect of the Assessment of Recruit Motivation and Strength (ARMS) Program on Army Accessions and Attrition*, Santa Monica, Calif.: RAND Corporation, TR975-A, 2011. As of July 27, 2018: https://www.rand.org/pubs/technical_reports/TR975.html

Marine Corps University Foundation, "University," undated. As of September 6, 2018: https://www.marinecorpsuniversityfoundation.org/university/

McFadden, Daniel L., "Econometric Analysis of Qualitative Response Models," in Zvi Griliches and Michael D. Intriligator, eds., *Handbook of Econometrics*, Vol. 2, Amsterdam: Elsevier, 1983, pp. 1395–1457.

Murray, Carla Tighe, senior analyst for military compensation and health care, Congressional Budget Office, *Evaluating Military Compensation, Statement of Carla Tighe Murray Before the Subcommittee on Personnel, Committee on Armed Services, United States Senate*, Washington, D.C.: Congressional Budget Office, April 28, 2010. As of February 10, 2018:
https://www.cbo.gov/sites/default/files/111th-congress-2009-2010/reports/04-28-militarypay.pdf

Nataraj, Shanthi, M. Wade Markel, Jaime L. Hastings, Eric V. Larson, Jill Luoto, Christopher E. Maerzluft, Craig A. Myatt, Bruce R. Orvis, Christina Panis, Michael Powell, Jose Rodriguez, and Tiffany Tsai, *Evaluating the Army's Ability to Regenerate: History and Future Options*, Santa Monica, Calif.: RAND Corporation, RR1637-A, 2017. As of May 19, 2017:
https://www.rand.org/pubs/research_reports/RR1637.html

National Center for Education Statistics, Institute of Education Sciences, "Recent High School Completers and Their Enrollment in 2-Year and 4-Year Colleges, by Sex: 1960 Through 2014," *Digest of Education Statistics*, Table 302.10, August 2015. As of February 16, 2018:
https://nces.ed.gov/programs/digest/d15/tables/dt15_302.10.asp

———, "Percentages of Persons 25 to 29 Years Old with Selected Levels of Educational Attainment, by Race/Ethnicity and Sex: Selected Years, 1920 Through 2016," *Digest of Education Statistics*, Table 104.20, November 2016. As of February 16, 2018:
https://nces.ed.gov/programs/digest/d16/tables/dt16_104.20.asp

———, "Number and Percentage Distribution of First-Time Postsecondary Students Starting at 2- and 4-Year Institutions During the 2011–12 Academic Year, by Attainment and Enrollment Status and Selected Characteristics: Spring 2014," *Digest of Education Statistics*, Table 326.50, February 2017. As of February 12, 2018:
https://nces.ed.gov/programs/digest/d16/tables/dt16_326.50.asp

Naval Postgraduate School, "Programs and Degrees," undated. As of July 31, 2018:
https://my.nps.edu/degree-programs

Nye, Christopher D., Fritz Dragow, Oleksandr S. Chernyshenko, Stephen Stark, U. Christean Kubisiak, Leonard A. White, and Irwin Jose, *Assessing the Tailored Adaptive Personality Assessment System (TAPAS) as an MOS Qualification Instrument*, Fort Belvoir, Va.: U.S. Army Research Institute for the Behavior and Social Sciences, Technical Report 1312, August 2012. As of May 25, 2017:
http://www.dtic.mil/dtic/tr/fulltext/u2/a566090.pdf

Office of People Analytics, U.S. Department of Defense, military enlistment processing station files provided to the authors, undated.

———, Status of Forces Survey of Active Duty Members results provided to the authors, August 2009.

———, Status of Forces Survey of Active Duty Members results provided to the authors, September 2016.

———, "Summer 2017 Propensity Update: Youth Poll Study Findings," February 2018.

Office of the Deputy Assistant Secretary of Defense for Military Community and Family Policy, *2015 Demographics: Profile of the Military Community*, c. 2016. As of August 2, 2018:
http://download.militaryonesource.mil/12038/MOS/Reports/2015-Demographics-Report.pdf

Office of the Under Secretary of Defense (Comptroller), *Military Personnel Programs (M-1): Department of Defense Budget—March Budget Amendment to the Fiscal Year 2017 President's Budget Request for BASE + Overseas Contingency Operations (OCO)*, March 2017. As of July 31, 2018:
http://comptroller.defense.gov/Portals/45/Documents/defbudget/fy2017/marchAmendment/fy2017_m1a.pdf

Office of the Under Secretary of Defense for Personnel and Readiness, "Annual Pay Adjustment," *Military Compensation*, undated. As of July 31, 2018:
http://militarypay.defense.gov/Pay/Basic-Pay/AnnualPayRaise/

———, *Report of the Ninth Quadrennial Review of Military Compensation*, Vols. I–V, Washington, D.C., March 2002. As of July 31, 2018:
http://militarypay.defense.gov/Portals/3/Documents/Reports/9th_QRMC_Report_Volumes_I_-_V.pdf

———, *Report of the Tenth Quadrennial Review of Military Compensation*, Vol. I: *Cash Compensation*, Washington, D.C., February 2008. As of August 28, 2018:
https://militarypay.defense.gov/Portals/3/Documents/Reports/10th_QRMC_2008_Vol_I_Cash_Compensation.pdf

———, *Population Representation in the Military Services: Fiscal Year 2010*, c. 2011. As of August 2, 2018:
https://prhome.defense.gov/M-RA/Inside-M-RA/MPP/Accession-Policy/Pop-Rep/2010/

———, *Report of the Eleventh Quadrennial Review of Military Compensation: Main Report*, Washington, D.C., June 2012a. As of July 31, 2018:
http://militarypay.defense.gov/Portals/3/Documents/Reports/11th_QRMC_Main_Report_FINAL.pdf?ver=2016-11-06-160559-590

———, *Report of the Eleventh Quadrennial Review of Military Compensation: Supporting Research Papers*, Washington, D.C., June 2012b. As of July 31, 2018:
http://militarypay.defense.gov/Portals/3/Documents/Reports/11th_QRMC_Supporting_Research_Papers_(932pp)_Linked.pdf

———, *Population Representation in the Military Services: Fiscal Year 2015*, c. 2016. As of August 2, 2018:
https://prhome.defense.gov/M-RA/Inside-M-RA/MPP/Accession-Policy/Pop-Rep/2015/

Orvis, Bruce R., Michael Childress, and J. Michael Polich, *Effect of Personnel Quality on the Performance of Patriot Air Defense System Operators*, Santa Monica, Calif.: RAND Corporation, R3901-A, 1992. As of November 23, 2017:
https://www.rand.org/pubs/reports/R3901.html

Orvis, Bruce R., Steven Garber, Philip Hall-Partyka, Christopher E. Maerzluft, and Tiffany Tsai, *Recruiting Strategies to Support the Army's All-Volunteer Force*, Santa Monica, Calif.: RAND Corporation, RR1211-A, 2016. As of November 23, 2017:
https://www.rand.org/pubs/research_reports/RR1211.html

Orvis, Bruce R., Christopher E. Maerzluft, Sung-Bou Kim, Michael G. Shanley, and Heather Krull, *Prospective Outcome Assessment for Alternative Recruit Selection Policies*, Santa Monica, Calif.: RAND Corporation, RR2267-A, 2018. As of August 28, 2018:
https://www.rand.org/pubs/research_reports/RR2267.html

Personnel Psychology, Vol. 43, No. 2, special issue: Project A: The U.S. Army Selection and Classification Project, 1990.

Public Law 110-252, Supplemental Appropriations Act, 2008, June 30, 2008. As of August 5, 2018:
https://www.gpo.gov/fdsys/pkg/PLAW-110publ252/content-detail.html

Rubenstein, Yona, and Yoram Weiss, "Post Schooling Wage Growth: Investment, Search and Learning," in Erik Hanushek and Finis Welch, eds., *Handbook of the Economics of Education*, Vol. 1, Amsterdam: Elsevier North Holland, 2007, pp. 1–68.

Schmitz, Edward J., and Michael J. Moskowitz, with David Gregory and David L. Reese, *Recruiting Budgets, Recruit Quality, and Enlisted Performance*, Alexandria, Va.: Center for Naval Analyses, CRM D0017035.A2/Final, February 2008. As of July 31, 2018:
https://www.cna.org/CNA_files/PDF/d0017035.a2.pdf

Scribner, Barry L., D. Alton Smith, Robert H. Baldwin, and Robert L. Phillips, "Are Smart Tankers Better? AFQT and Military A Productivity," *Armed Forces and Society*, Vol. 12, No. 2, 1986, pp. 193–206.

Segall, Daniel O., *Development and Evaluation of the 1997 ASVAB Score Scale*, Defense Manpower Data Center, July 2004. As of July 27, 2018:
http://official-asvab.com/docs/1997score_scale.pdf

Sellman, W. Steven, "Predicting Readiness for Military Service: How Enlistment Standards Are Established," draft prepared for the National Assessment Governing Board, September 30, 2004. As of August 1, 2018:
https://www.nagb.gov/content/nagb/assets/documents/what-we-do/sellman.doc

Simon, Curtis J., and John T. Warner, "Managing the All-Volunteer Force in Time of War," *Economics of Peace and Security Journal*, Vol. 2, No. 1, January 2007, pp. 20–29.

Stark, Stephen, Oleksandr S. Chernyshenko, Fritz Drasgow, Christopher D. Nye, Leonard A. White, Tonia Heffner, and William L. Farmer, "From ABLE to TAPAS: A New Generation of Personality Tests to Support Military Selection and Classification Decisions," *Military Psychology*, Vol. 26, No. 3, 2014, pp. 153–164.

U.S. Army War College, "Military Education Level 1 Programs," undated. As of July 31, 2018:
https://www.armywarcollege.edu/programs/mel_1.cfm

U.S. Census Bureau, "Current Population Survey (CPS): Data," last updated March 6, 2018. As of August 31, 2018:
https://www.census.gov/programs-surveys/cps/data-detail.html

U.S. Code, Title 26, Internal Revenue Code; Subtitle C, Employment Taxes; Chapter 21, Federal Insurance Contributions Act. As of August 2, 2018:
https://www.gpo.gov/fdsys/granule/USCODE-2011-title26/USCODE-2011-title26-subtitleC-chap21/content-detail.html

U.S. Department of Defense, *Career Compensation for the Uniformed Forces: Army, Navy, Air Force, Marine Corps, Coast Guard, Coast and Geodetic Survey, Public Health Service: A Report and Recommendation for the Secretary of Defense by the Advisory Commission on Service Pay*, Washington, D.C., 1948.

———, "Fall 2016 Propensity Update, Youth Propensity: Youth Poll Study Findings," Joint Advertising and Marketing Research and Studies, Office of People Analytics, 2017, unpublished.

U.S. Government Accountability Office, *Military Personnel: Military and Civilian Pay Comparisons Present Challenges and Are One of Many Tools in Assessing Compensation*, Washington, D.C., GAO-10-561R, April 1, 2010. As of August 30, 2018:
https://www.gao.gov/products/GAO-10-561R

U.S. Naval War College, "Programs Offered," undated. As of July 31, 2018:
https://usnwc.edu/Academics-and-Programs/Programs-Offered

Warner, John, Curtis Simon, and Deborah Payne, "The Military Recruiting Productivity Slowdown: The Roles of Resources, Opportunity Cost and the Tastes of Youth," *Defence and Peace Economics*, Vol. 14, No. 5, October 2003, pp. 329–342.

Wenger, Jennie W., Zachary T. Miller, and Seema Sayala, *Recruiting in the 21st Century: Technical Aptitude and the Navy's Requirements*, Washington, D.C.: CNA Analysis and Solutions, CRM D0022305.A2/Final, May 2010. As of May 19, 2017:
https://www.cna.org/CNA_files/PDF/D0022305.A2.pdf

Winkler, John D., Judith C. Fernandez, and J. Michael Polich, *Effect of Aptitude on the Performance of Army Communications Operators*, Santa Monica, Calif.: RAND Corporation, R4143-A, 1992. As of November 23, 2017:
https://www.rand.org/pubs/reports/R4143.html

Yule, G. Udny, "Why Do We Sometimes Get Nonsense-Correlations Between Time-Series? A Study in Sampling and the Nature of Time-Series," *Journal of the Royal Statistical Society*, Vol. 89, No. 1, January 1926, pp. 1–63.